SAFETY METRICS FOR THE MODERN SAFETY PROFESSIONAL

SAFETY METRICS FOR THE MODERN SAFETY PROFESSIONAL

C. Gary Lopez

CRC Press
Taylor & Francis Group
Boca Raton London New York

CRC Press is an imprint of the
Taylor & Francis Group, an **informa** business

First edition published 2021
by CRC Press
6000 Broken Sound Parkway NW, Suite 300, Boca Raton, FL 33487-2742

and by CRC Press
2 Park Square, Milton Park, Abingdon, Oxon, OX14 4RN

Library of Congress Cataloging-in-Publication Data

Names: Lopez, Gary, author.
Title: Safety metrics for the modern safety professional / Gary C. Lopez.
Description: First edition. | Boca Raton, FL : CRC Press, 2021. | Includes bibliographical references and index.
Identifiers: LCCN 2020026903 (print) | LCCN 2020026904 (ebook) | ISBN 9780367497057 (hardback) | ISBN 9781003088332 (ebook)
Subjects: LCSH: Industrial safety--Evaluation--Statistical methods. | Industrial management--Evaluation--Statistical methods.
Classification: LCC T55.3.S72 L67 2021 (print) | LCC T55.3.S72 (ebook) | DDC 363.11072/7--dc23
LC record available at https://lccn.loc.gov/2020026903
LC ebook record available at https://lccn.loc.gov/2020026904

ISBN 13: 978-0-367-49705-7 (hbk)

Typeset in Times
by MPS Limited, Dehradun

**Visit the Taylor & Francis Web site at
http://www.taylorandfrancis.com**

**and the CRC Press Web site at
http://www.crcpress.com**

Contents

Preface

I began my journey in the field of safety in 1976. That seems like an eon ago as I look back at the long journey on which my career path has taken me. As I reflect on how the field of safety has evolved, as well as the safety professionals in that field, this journey might as well háve started in the last ice age.

Speaking of reflection, as I think back regarding the metrics I was "forced" to use as I entered the field (and for some time after), I am ashamed to say that I did not comprehend the shortcomings of these metrics more quickly. However, like many other safety professionals of the time, I "inherited" them and did not question them. After all, those metrics had been in use for over half a century. Who was I to question them? To not use them would have made me look like a heretic.

Perhaps with expanded responsibility, comes expanded vision. As I rose up the career ladder in my company, I found that these metrics were inadequate. They simply did not measure up (not to make a pun) to address the scope of what I did in terms of managing the risks for my organization. In fact, to limit me to these "employee injury only" metrics would have limited my role regarding the types of risks I was managing. They were narrow in focus, and, worse, the numbers used were abstract and little understood by my peers on the management team. In fact, once you left the circle of my fellow safety professionals, these numbers made little sense.

Even this latter assumption of my peers in the safety field understanding them is probably an exaggeration. As I discussed this dilemma with many of those peers, I found that, in many instances, they did not dwell on these numbers or if they truly measured all we did. Many could neither identify where the formulas came from, nor the logic behind them. We simply echoed and repeated them like trained parrots.

Admittedly, this personal "metric crisis" was of my own doing. Your view from the top of the corporate ladder changes how you see things as compared to when you showed up for your first day of work as a safety engineer (which was my first title). As you rise to the top of the corporate ladder and deal with others who are taking a strategic look at what they are doing, it causes even the most apathetic thinker to take an introspective look at if what you are doing brings true value to your organization. It compels you to ask introspective questions, such as am I measuring success (and conversely failure) well? Am I grasping our strategic mission to my organization? Am I worth the handsome salary my company pays me? Or, put another way, "What am I doing to earn my paycheck?"

These questions led me to understand that my job was larger in scope than managing employee injuries and illnesses. Clearly, I was managing risk to the organization. A much bigger picture. While this kind of thinking is becoming commonplace today, it was pure heresy in the "zero accident" era of the 1980s when I began this journey. Nevertheless, it led me down the path of understanding that measuring success in my efforts was going to require more sophisticated approaches to metrics than lost time accident rates or recordable injury rates.

This metric crisis I was going through was only exacerbated when I became part of the business team. Now I was the one who had to ask for capital dollars for safety improvements or design changes. While other members of the business team used the dollar as a metric to justify their requests, I was using no logic other than "because it is safety" as my rationale. Soon I had to learn to use the dollar as my logic.

In the following pages, I will take the reader through the journey I took in expanding metrics to identify success in what we do for our organization. While each organization's goals and objectives may demand various metrics, the concept of measuring this success in more than one area, or should I say dimension, of what we do as safety professionals is a constant.

My major purpose in writing this book was to pass this knowledge along to future safety professionals and those struggling with some of the same measurement issues I experienced in my career. It is my hope that this will provide some guidance and insight into what the 21st-century safety professional will need in their journey through their career.

I would also be remiss in not pointing out to the reader that what you read in the following pages is not "academic theory." This is not meant as an aspersion of academia in any way. Such "theory" can come not just from higher education, but also from someone having a flash of brilliance or sincerely believing their great idea will work. By "experience," I mean to say that what the reader will see in the following pages is what I saw work well in my organization, not my flashes of brilliance, of which there were few. This should comfort the reader in knowing that the examples I provide in the following pages are not based on the latest "safety craze" or fad. What you will read about in these pages is knowledge gained through hard experience. Some knowledge, I might add, painfully gained the hard way. By making mistakes. Sometimes the hardest lessons in life are the best lessons in life. We learn from our failures as much as we do from our successes. Perhaps more so.

That said, the reader will find that a continual theme in the ensuing pages is one of the flexibility and customization of your metrics. There is no magic pixie dust when it comes to developing metrics for an organization. Learn from my experience from the past with one eye to the future.

As a final note, I must make some apologies upfront for many of the examples and terminology I use in this book. They are obviously biased in most instances toward the "for-profit" business sector in which I spent my career as well as the era I lived in. I plead guilty to all charges. I spent the entire "first phase" of my working career (before flunking at retirement) in the manufacturing sector. As I shared with the reader earlier, this book is based on experiences and not theory. My experiences were in a large multinational chemical/pharmaceutical organization. Consequently, the examples I cite in the following pages are biased to that experience.

Which brings up the use of the word "I" in much of the document. I will be using the pronoun "I" quite a bit in discussing the various examples and results. In fact, "I" will use it even more than I am comfortable with using that particular pronoun. One of the great lessons I learned in management was that teamwork is a path to success. I noticed that the more I surrounded myself with competent smart people, the more I got "competent and smart." Nevertheless, as I stated, this is a book based on experience and not theory. Since the experiences were mine, you will have to bear with my continual use of this pronoun in many of the examples and scenarios I cite. Try to move past that and realize that it was the team that contributed to my successes, not the captain of the ship.

As for some of those biases, perhaps apologies are not necessary. Since my retirement from the industrial sector, I have embarked on a "second phase" of my career. In this second phase, I have been exposed to many different types of organizations including government, education, construction, nonprofits, and my old friend manufacturing. This experience has verified what I suspected. Every organization, no matter what type, struggles with metrics to measure the success of their safety programs. The metrics issues and solutions I suggest in the following pages are very much applicable across the board.

While I am on apologies, I will also apologize for any perceived insensitivity to females in our workforce. When I use terms such as "man-hours," it is not meant to be sexist. Perhaps a more politically correct term for the present would be "service hours or employee hours." The context in which I use the term is more historical in nature. It is simply to convey representation of how things were measured in the past. In my defense, to this day I view the term the same as if I were to say "you guys" to a mixed group of men and women friends (where I grew up this is still a common phrase which is much like y'all in the south). This apology also extends to many of my examples in which you will find me using "he" or other male-based descriptors for much of the upper management I dealt with in the past. It was simply the state of things at the time. Upper management was a very male-dominated field. At any rate, I hope this suffices to deflect any "gender" criticism in the following pages.

With apologies covered and rationale for the book explained, I will desist in any further introduction and move on. I sincerely hope that the reader will gain some knowledge in the following pages about new approaches to applying metrics in your organization.

I have long said that if you gain one *new* idea by reading a book, attending a seminar, or any other means of education, the endeavor was well worth it. Hopefully, the reader will gain at least one new idea or concept from the following pages.

C. Gary Lopez
Weston, Florida

Author

 **C. Gary Lopez, MS, CSP, FASSP,** is an accomplished safety professional with over 40 years of safety and risk management experience in the manufacturing, construction, and insurance industries.

Lopez has spent the majority of his career in the chemical and pharmaceutical industries. He has served in a variety of management positions starting out at the plant level and working his way up to the corporate executive level. He has experience in operations safety, prevention through design, and instituting safety management systems in the chemical and pharma industries. As part of his executive duties, Lopez was one of the first to combine the safety, security, insurance, and industrial hygiene functions into one department that took a holistic view of managing risk for the organization. As part of these responsibilities, he recognized the need to develop metrics systems that measured the success of safety and risk performance in all areas of risk within the locations he was responsible for in his position.

Lopez is a graduate of West Virginia University with a master's degree in safety. He is a Certified Safety Professional, a Fellow of the American Society of Safety Professionals (ASSP), a recipient of two Culbertson Awards from the ASSP, as well as a recipient of the Bresnahan Medal for standards development. Lopez is also a recipient of the National Safety Council's highest award, the Distinguished Service to Safety (DSS) Award. He also chaired the Chemical Section Executive Committee when they won the Cameron Award.

In addition to these accomplishments, Lopez was heavily involved in the Chemical Manufacturer's Association (now the American Chemical Council) development of the Responsible Care™ Program, chairing the committee that drafted the Safety and Health Code and the CAER Exercise Guide.

Currently, Lopez is President of Risky Biz Services Inc. and is also employed by Gallagher Risk Management Services; and he resides in Weston, Florida.

1 We Manage What We Measure

We Manage What We Measure.

C. Gary Lopez

Perhaps one of the most-often used paraphrased quotes about metrics in business is attributed to Peter Drucker. Why do I say paraphrased? Unfortunately, for Peter, his original quote, which made an excellent point regarding all metrics in use in any management system, is also one of the most misquoted of his quotes. His exact quote was "If you can't measure it, you can't improve it." Over the years this got paraphrased into "You can't manage what you can't measure," or "You manage what you measure." I will join that long list of those paraphrasing the quote with "We Manage What We Measure."

I submit that the "misquotes" are more catchy. More to the point, they all illustrate the same issue. There is a simple truth in the fact that if we don't devise a system of measuring something we certainly have a difficult time demonstrating we are improving or getting worse. We truly manage what we measure.

I am a golfer. Fortunately (or perhaps, unfortunately) for me, a system of measuring golf performance was devised long before I began to play. Each hole is clearly identified regarding what constitutes par. I can rant and rave about how fair it is, but, at the end of the day, par is par. Everyone else on the course is dealing with the same scoring system. This scoring system, or metric, allows me to compare my performance to others on the course.

You can go above par, by scoring a bogie (or worse) or you go below par with a birdie or eagle. The point is that the scoring system for every course, no matter how different the course layout is, is identical. Course par may vary course to course, but the "system" is the same. No matter where I go in the world, every golf course uses this system. It is a simple system every golfer in the world can use and understand. I can play a course that Tiger Woods plays and know how I performed according to "par for the course."

Furthermore, in order to make the scoring system even more flexible, we also have a "handicap" system for adjusting your score. By using this measurement system, I can calculate my handicap and adjust my score to level the playing field against better players. As an example, if by chance Tiger wants to play against me (an eventuality we will probably not have to worry about), he would give me strokes based on the fact his handicap (and talent level) is considerably better

1

than mine. I can now step on the first tee knowing that despite this difference in talent, I have leveled the playing field—*Theoretically.*

The end product of this elaborate scoring system is that it answers the most basic of golf questions. "Am I improving, standing still, or declining in performance?" If only we had it so easy in the safety profession measuring our "scores."

I can think of no greater challenge that the modern safety professional faces than to accurately measure are we improving, standing still, or declining in performance. Am I shooting par for the course? Or am I doing much worse? We need a scoring system that identifies success.

Exacerbating the difficulty of our challenge is the nature of our greatest "successes" in our chosen field. Allow me to elaborate.

Several years ago, the American Society of Safety Professionals (then the ASSE) honored me as a Fellow of the Society. In order to be accorded this honor, you must dig into your past and demonstrate what you have done in the field of safety to deserve such an honor. In my case, that meant bringing dusty files down from the attic, tracing my history and accomplishments during my career in safety.

At the time you had to accumulate all of this data in hardcopy form. You then sent several copies of the books to the honors committee for review. As I stared at the volume of my life's work, I could not help but speculate on a simple fact. A fact that I shared with the audience when I conveyed my acceptance speech.

As I stood on stage, I shared with the audience my walk down memory lane in documenting my career. Then I pointed out a fact that all past and current safety professionals will spend their careers experiencing. That fact is that we will never know what accidents we prevented by our actions. We will never know what employees went home unharmed, what factories did not blow up, what buildings did not catch on fire, and what other terrible disasters we spared our organizations because of our efforts. We are not salespeople who can proudly point to what we sold. We are in a profession in which our greatest successes were what did not happen.

However, while this may be a true statement, it does not let us off the hook. We cannot march into the CEO's office and say "I want a huge bonus this year because 'theoretically' I saved several lives, a plant from exploding, and bad press for the company."

It might be worth a try, but I would not advise it. Instead, we are presented with the problem of measuring what did happen more accurately and perhaps what measurements can be agreed upon by all parties that identify meeting the goals of the organization. This makes the choice of metrics we use to measure the success, or failure, of our efforts that much more difficult. It also means that to more accurately reflect our successes, we must expand what we have measured in the past.

In accomplishing this goal, the reader may find (much to your pleasure, or perhaps displeasure) that I am not going to suggest throwing out what we have come to call "lagging indicators." I know it is popular now to criticize lagging indicators, but they are still one of our most powerful tools in not just demonstrating loss, but in justifying expenditures in the future to put controls in place to address such loss. In fact, I am going to promote the use of more lagging indicators, not less.

While there has been much written lately about how these indicators measure failure, not success, I have a different view. They are simply one more measure we have in our toolbox that we need to use wisely. As an example, every company has a profit and loss statement. This is a lagging indicator, but it is not viewed in a negative manner. It is a simple reporting of results. Our lagging indicators should be no different. However, like the P&L statements, what we report should be meaningful and accurate.

For the past decade, I have been on a mission to see the old ANSI Z16.1 standard brought back to life, but with a whole new "look" that identifies what metrics the 21st-century safety professional will need to more accurately reflect their mission. For those of you not familiar with the ANSI Z16 standard, until the advent of OSHA, it was the gospel on how we recorded and measured workplace injuries. In fact, the name of the standard was *Recording and Measuring Work Injury Experience*.

Many younger professionals have probably never heard of it. There is a good reason for this. I will cover this in greater depth in the next chapter, but essentially once OSHA came into being, they decided, for reasons unknown to this day, to not adopt this particular ANSI standard and created their own recordkeeping system using the Bureau of Labor Statistics. Slowly, but surely, more and more organizations abandoned the ANSI Z16.1 system and switched over to the OSHA recordkeeping system as their metric.

As time marched on, it became increasingly apparent that this system became the predominant measure of safety success or failure in our field. However, problems did arise in applying the system at the microlevel. While it might seem like a simple task to convince everyone that the OSHA system just didn't work that well in telling the tale of what we do as a profession that was hardly the case. Changing metrics in our profession is no small feat.

However, as one who had been pushing for such a change, I decided we had to start somewhere. I choose to start with the now-defunct ANSI Z16.1 to "sell" our new metrics. Why start with a defunct standard, I will get to that in a moment. But first, I had to sell the idea to ASSP to serve as the secretariat for the standard. Once that sale was made, I had to sell the concept to potential committee members. Some remembered the old Z16.1 and were not clear on why we were going to bring it back from the dead. I had to convince the new committee that this was not to be a regurgitation of the old Z16.1 with modern buzzwords.

I would like to provide the reader some insight into why I saw ANSI as the best mechanism for this mission. All ANSI standards have a simple provision that keeps them as "state of the art" for best practices. While OSHA has elaborate and ponderous steps, it must go through to change anything, ANSI is much more flexible and expeditious when it comes to change. By ANSI requirements, every ANSI standard must do the following every five years:

- Reaffirm the standard
- Revise the standard
- Sunset the standard

The obvious upside to this is that you cannot write an ANSI standard and freeze it in time for 40 years. Despite OSHA's best efforts, they have difficulty changing standards they know are out of date. In the case of ANSI, they are "forcing" all of their standards committees to keep current by either stating it is still state of the art, it needs updating or it is no longer necessary so should be withdrawn.

This ANSI system offers another advantage. The operative word when dealing with safety metrics in the 21st century will be "change." There will be a constant need to address the rapid change taking place in the field and meeting the new requirements for measuring safety success. What better system for this than ANSI.

The intent of the original committee back in the 1920s was to create the first system for measuring safety injuries. This new ANSI committee would be the first to measure total risk.

Does this then start with an assumption that the new committee won't get it right the first time? Absolutely. I have learned in my working career that one thing that is inevitable is change. It would be complete arrogance to think that no matter who was on the committee writing this revolutionary new standard that no further change would be necessary. Quite to the contrary, as soon as the standard is published, there is almost a guarantee that in five years the next committee will improve it. That is why the ANSI system was selected for this endeavor.

As part of this effort, one of the first people I reached out to was Mr. Steve Newell, a principal partner in ORCHSE. Why Steve? Like me, Steve has spent a great part of his career in search of the "holy grail" of meaningful metrics to our profession. Although admittedly Steve's journey has been a bit more "colorful" than mine and certainly one that had him more in the spotlight.

I first met Steve when we worked as part of the "Keystone Group" many years ago. This group was tasked with trying to make OSHA recordkeeping requirements, then handled by the Bureau of Labor Statistics (BLS) more intuitive and sensible. Since I worked for a chemical company, I was nominated through the Chemical Manufacturer's Association (CMA) to represent chemical industry concerns about recordkeeping practices.

The Keystone Group was a mediation effort using an organization, Keystone Group International (hence the nickname for the "Group"), as a mediator to manage the process and the various entities such as labor unions, industry groups, government, etc., to agree on an improved process.

The need for this clarification came when OSHA launched one of their first special emphasis like programs targeting recordkeeping violations. Through a series of high-profile inspections, OSHA began to issue a plethora of citations dealing with recordkeeping.

What made these inspections unusual was that instead of simply issuing one citation for improper recordkeeping, as they had done in the past, OSHA issued a citation for each instance of improper recordkeeping thereby driving up the fines. This got the attention of a number of industries. Fines that had once been a nuance were now becoming pricey.

The issue that everyone agreed was paramount in resolving was the clarity of what was and was not recordable according to "The Blue Book." For those around in that period, the "manual" for assisting in determining recordability was a question-and-answer format book that was issued by BLS to explain recordkeeping. Since the book was blue in color, it received the Nom de Guerre of "The Blue Book." Steve had the distinction of being the architect of this Bureau of Labor Statistics famous, or, depending where you stood on the topic, infamous Blue Book.

In Steve's defense, he did not foresee that most of the safety profession would seize these measures and go right down to the department level using them, despite the fact they were a macrostatistic. Consequently, the OSHA recordkeeping system of incident rates, lost time accident rates, and limited duty cases would take on a life of its own.

I would like to take a moment to jump to Steve's defense for his role in the entire Blue Book episode. When I first read the Blue Book, I was not among his legion of admirers (he would say there was no legion, more like a small gang of admirers if that). However, once I got to know Steve, I realized he was not some bureaucrat who wrote the book in isolation and with no knowledge of the topic. For the most part, he was a thoughtful and intelligent guy who was swept up in the tidal wave of OSHA recordkeeping that spiraled out of control. In fact, during the meetings with the Keystone Group, he was one of the most logical and sensible for coming up with solutions to clarifying the recordkeeping requirements. Nevertheless, to this day, he cringes when anyone brings up the topic of the Blue Book.

When I was selected as chair for the new ANSI Z16.1 committee, the first person who entered my mind to become vice chair was Steve. To say he was skeptical of the whole idea is putting it mildly. He, like I, had struggled for decades trying to get his arms around the metrics issue and he did not see a standard as the cure-all. Consequently, after selling the concept to the ASSP, he was my second "sell."

I presented a case to Steve for what we were trying to do. As Steve put it, I had laid out the sale in "four buckets." I give him full credit for that concept and have been using the "bucket" approach to describe what the intent of the new standard was ever since.

The Four "Buckets" were as follows:

Bucket #1 Traditional Lagging Indicators

> Clearly, traditional lagging indicators cannot be thrown out. Considering that we have 100 years of workplace accident data kept in this format, there will still be a need for tracking and using this type of metric as a measure. However, simply using employee injuries and illnesses as the single lagging indicator is a one-dimensional measure of safety in the workplace, not the final definitive measure.
>
> Furthermore, for US and international companies with US-based locations, there is a need to meet the recordkeeping requirements of OSHA and other regulatory agencies internationally.
>
> Perhaps more to the point, lagging indicators can play an important role in supporting the success of leading indicators and also provide a concise picture of the success of the safety efforts depending on how they are structured.
>
> It is the form these lagging indicators will take that is essential. The form these lagging indicators take will change dramatically from what has been used in the past. Lagging indicators have not been the problem. It was the lack of imagination in how we used these indicators that cast a negative light on them.

Bucket #2 Leading Indicators

For years, there has been a movement to shift from simply tracking failures to more positive management metrics. To some extent, the over reliance on OSHA rates has blocked movement to the use of more proactive systems of measuring risk. Leading indicators have the potential to provide critically needed management tools that can be used in prevention efforts.

The concept behind leading indicators is to first assess what risks an organization face and then to determine what action items can be carried out by supervision to address these risks. Or looked at in another way, what interventions or controls can be put in place to address the risks.

This "leading" approach not only gives supervision a solid understanding of their role in workplace safety, but also provides metrics that can be measured and input into supervision performance reviews.

Although called "predictive," these leading indicators are more a matter of determining what actions would impact the risk of an organization and seeing that they are carried out and measured.

Leading indicators are the gateway to providing management with demonstrable things they can do. In this role, they can play a powerful role in enhancing an organization's safety culture and providing a metric for our profession.

Bucket #3 Value-Based Metrics

No metric developed in the last 100 years will mean more to the safety professional than value-based metrics. This is the future. Safety and health professionals have historically focused on employee injuries and illnesses and left the rest of the "losses" to the insurance or other departments. Areas of risk such as fleet, property, general liability, business interruption, and security have normally been left to other departments in an organization because safety professionals allowed their metric of choice to narrow their view of their true role of what they bring to the table in any organization. The outcome has been a tunnel vision safety professional that sees their role as just preventing employee injuries.

This "bucket" will depart from that tradition. The first departure is that the primary method of tracking loss will be the international language of business; the dollar sign. Value-based metrics are intended to drive safety professionals to begin to quantify the metrics they choose. Whether it be using them to determine a return on investment or how to accurately measure the true cost of an accident, these dollar signs will serve as the most overt metric our profession can use. Metrics that will go beyond the traditional insurance loss run for workers' compensation. It will bring the concepts of Acceptable Risk and Residual Risk into the metric process. In short, it will force our field to think like businesspeople as well as safety professionals. It will bring clarity to our mission and make us fit in more with other managers on the business team.

Bucket #4 Engaging Senior Business Leaders

Safety and Health professionals seem to be challenged when it comes to engaging senior leaders in setting what metrics define success. Most safety professionals are more comfortable thinking in technical terms that make sense to us, but often do not resonate with the rest of the business. This bucket will provide guidance to break out of this cocoon.

Upper management needs guidance on how to set the tone for the safety culture of an organization and what metrics it wants to set as values for an organization. The future safety professional's job is to sit down with upper management and provide them with guidance on their options.

Different executives hold different values and have different priorities for their organization. All CEOs will say they want to have a safe company, but what does that mean? Values have to be set and the safety professional should be assisting in suggesting those values and then having the CEO endorse them.

Engaging upper management in identifying what constitutes success must be done at all levels in an organization. Further to this point, what you might measure at the upper management corporate level will not necessarily be what you would measure at the department level. However, the measurements must complement each other and not diverge from a common definition of success.

As an example, what I would measure at the corporate level might be three or four key performance indicators (KPIs). At the middle and frontline levels of management, you might measure a mix of a dozen items mixing leading indicators with lagging indicators. However, all of these would complement the upper level KPIs. They would all be interlocked into one set of metrics. The metrics that identify success.

DEFINING SUCCESS

In the following pages, I will spill the contents of these four buckets into each of the chapters. I will elaborate on what these contents mean to the safety profession and to those to whom we report our statistics.

We must never lose sight of the fact that metrics are powerful tools that tell the story of what we do as a profession and how we identify success in what we do. In the following pages, you will see examples of many forms of metrics and measures that can be used. They will be in a variety of formats and values.

Our job as safety professionals is to work with our management teams to agree on which of these formats and measures define success in the eyes of the organization for which you work. They will not all be the same because every organization has a different set of values – values that are shifting not just organizationally but globally.

REFERENCES

Peter F. Drucker. *Management; Tasks, Responsibilities, Practices*. Harper & Rowe.
ANSI Z16.1. 1954. *Recording and Measuring Work Injury Experience*. American National Standards Institute.
Injury and Illness Reporting Guidelines (Blue Book). Bureau of Labor Statistics.

2 A History of 20th-Century Safety Metrics

Those who do not learn history are doomed to repeat it.

George Santayana

Before we proceed any further, I believe it is important to understand some basic history on safety statistics and measurements in order to understand how we got to where we are currently.

In majority of the safety presentations I conduct, I always attempt to include a "history" lesson in the topic matter. Why? Because we are a profession of historians. At least if we approach our field wisely, we should think of ourselves as historians.

This epiphany first dawned on me when I was in college, studying for my master's degree in safety. We were assigned to do a thesis, preferably on a significant accident that had occurred in history and what we learned from it. Most of my counterparts chose the Triangle Shirt Factory or Coconut Grove fires. Both of these events had a considerable impact on fire protection in our country. One could posit that they were responsible for creating the National Fire Protection Association (NFPA) we know today, an association that has extended its reach into the fire codes of every community in this country. Although I found those incidents to be thought provoking, I decided to go back a bit further in time. I chose the sinking of the *Titanic.*

As I dove into my topic, it was an illuminating experience. For the younger reader, this was before James Cameron's famous movie; so when I say "illuminating," it has nothing to do with the fate of Jack and Rose. It had more to do with how the failure of both engineering and administrative controls piled one on the top of another to create a disaster of unimaginable loss.

Research was not easy back then. There was no Internet to turn to for information. I could not simply "Google" it and have the facts of the accident pop up on my computer screen. I had to dig. I had to dig in libraries for books on the topic, reports of the investigations that followed the disaster and even old newspaper articles.

The more I dug the more fascinated I became with what happened on that icy night of April 15, 1912. The more I dug the more I learned of all the things that could have been done to prevent it from happening or steps that could have been taken to lessen the loss. It was a complete lesson in engineering failures, communication breakdowns, management structure dysfunction, and all the components that can come together to make a great disaster. In summary, it was the "perfect storm" for a disaster of unimaginable magnitude.

To put perspective on how shocking this disaster was at the time, try to imagine how our modern news media would go crazy if the largest cruise ship on the planet, an "unsinkable" ship on her maiden voyage, was lost and then tack on a passenger list that included Warren Buffet, Bill Gates, Elon Musk, and Steve Jobs. That is a modern rough equivalent to the list of wealthy passengers on the passenger manifest that night. Nevertheless, even if it had not been the largest most luxurious ship of its time with an unheard of list of wealthy passengers, it was a compelling disaster to study.

Now, some of you may be speculating right now, "What does a shipwreck in 1912 have to do with what we are dealing with in the 'modern' world?"

I led this chapter with George Santayana's famous quote since I have found it to be true many times in my career. *"Those who do not remember the past are condemned to repeat it."* For our profession, this quote should rest on our desks in front of us, or on a wall plaque that we pass every day. We need to be aware that if we neglect our past accident history, we could find ourselves reliving it.

This is perhaps what hooked me on the *Titanic* incident. It was more than a simple lesson of a past accident. What stuck with me was that everything that happened that night of April 15, 1912 could happen again in a different scenario. The scenario may not have been identical. But the "failures" in communications, good emergency response planning, design, and management miscommunication that night were all basic principles that could resurface in almost any industry.

I was so taken by this revelation that when I was hired on my first job as a young safety engineer, the first purchase I made was a picture of the *Titanic*, which I hung behind my desk. That picture traveled with me to every office I occupied on my corporate ladder climb until I retired. I must admit it made a few of my bosses and the executives who venturred into my office nervous when the first thing to greet them visually was one of the great disasters of history. More than one commented about their "safety guy" having a picture of the *Titanic* hovering in the background not exactly instilling confidence in the troops. However, I assured them it was there as a reminder for me not to ever get overconfident. We must never think we have an "unsinkable ship."

So what does this have to do with metrics you may be asking at this point?

As I stated earlier, we are a profession that is cursed by some extent by our successes. We cannot measure what did not happen. Consequently, we need to learn more from what did happen. As I pictured myself as the safety director of the White Star Line after this incident, I certainly could not approach our board or the investigators and measure the impact of this disaster just by tracking the employees killed and injured. My metrics would have to be much broader than that to capture the magnitude of the loss. Such thinking started me down the road of using value based metrics to track total loss, not just employee injuries. Even if it had been invented back then, the LTA rate was hardly sufficient to place a measurement on the amount of this loss.

Furthermore as I pondered the hearings that were held to investigate this disaster my mind flashed back to the impact these hearings would have on future design and operations not just of the White Star Line ships (the owners of the *Titanic*) but all other shipping companies. As an example, never again would a ship sail without enough lifeboat seats for its passengers. Never again would ships not have radio sets manned around the clock to take in SOS signals.

As I imagined myself dealing with this scenario in 21st-century terms, I would not get away with simply reporting what the losses were in terms of workers, compensation numbers. I would have to face the total cost of the incident. Using the cost of accidents that happened to others can clearly illustrate what the potential cost could be to your organization if you have a similar exposure. Consequently, when working for a competitor such as Cunard Line, if I was calculating the return on investment (ROI) on investing in better lifeboat capacity, I am no longer pulling numbers out of the air. Transferring this logic into our modern world I have real experience to base my calculations on. I do not need a tank farm fire to occur in my company to justify safety measures. All I have to do is add up the price of a similar fire in another company and I will have my potential costs. As historians, we can learn from the past.

There is an old "tongue-in-cheek" saying in the safety profession; "There is nothing better for the safety business than a disastrous accident; as long as it happens to the other guy." Certainly, there is more than a grain of truth in this statement. I personally witnessed an increased safety awareness from management after other companies suffered serious accidents. It did not take long for the "could this happend to us" queries to hit my desk. We can use such metrics of loss to our benefit. However, waiting around for a disaster to hit the other guy is not necessarily the best strategy. Developing our own value based metrics makes far more sense.

However, this change in expanding our metrics will not come about immediately. Such thinking will take radical change from our profession since we seem to be stuck on the same numbers we have been using for years. When it comes to the topic of metrics, most safety professionals hardly give the history of where they came from a second thought. Nor that they are in need of change. I cannot be too critical of this stance.

Admittedly, I suffered from it myself for years. When I got my first job, I was told what we measured and I never questioned it. Essentially, what they were saying to me was "Here is what the guy before you measured, so you do the same thing." Don't ask why. Don't ponder on why we measure it. Don't ask if it really tells the story of what we are doing or what we should be doing. Just use it.

I certainly was not alone in this introduction to safety metrics. In fact, many in our profession can go through an entire career, not questioning the metrics or pondering if they are sufficient. Why not? Our profession has spent 100 years selling these numbers. It has been drilled into most executive management's heads by us for years as "the number." Who are we to question it?

Although much of this logic is true, this does not mean they are the right numbers or that we should not question them or improve them. However, this chapter is about history, so let us examine where the numbers came from so that we can better understand them.

BIRTH OF A RECORDKEEPING STANDARD

As we journey back in time to see how our statistics originated, we find that it was around the turn of the 20th century that employers started to become concerned about employee injuries. I would like to say that it was an altruistic movement on the part of executives in American industry that generated this concern. However, in truth, the major compelling reason was the oldest of reasons: money.

Shortly after the turn of the 20th century, both federal and state governments began to enact laws that were providing fair compensation for injured workers. It was becoming a societal issue that industry could not ignore. The days of looking at an injured worker as a broken part to be discarded were coming to an end.

The first workers' compensation laws were being passed by states around this time. As one might expect, these laws were adding a new operating cost that had to be addressed. Consequently, companies began to pay attention to the safety of their workplaces to control these rising costs.

This was exacerbated by the fact that insurance companies began to place pressure on businessmen who wanted to purchase "worker injury" insurance for their employee injuries. Insurance companies were forcing those who insured to put controls in place to lessen the chance of accidents. In short, this was the beginning of the safety movement as we know it in this country.

Soon two organizations were formed that would begin to shape the field of safety. As a note, both of these organizations would owe their birth in a large part to disasters.

On March 25, 1911, the Triangle Shirtwaist Factory fire occurred. This tragedy resulted in the forming of the American Society of Safety Engineers (ASSE) in that same year (now the American Society of Safety Professionals). Shortly thereafter, in 1913, the National Safety Council (NSC) was formed to address safety issues in our country (technically, it did not get that name until 1914; originally, it was called the National Council for Industrial Safety). The first steps in the "professionalizing" of our safety profession were taken in the journey toward professionalization.

In 1912, the Bureau of Labor Statistics (BLS) decided to conduct a study of injuries in the iron and steel industries (of note that same year the Association of Iron and Steel Electrical Engineers sponsored the first Cooperative Safety Congress which later led to the formation of the NSC).

As BLS began its study on workplace injuries, it became immediately apparent, what they did not have was a common way of measuring these injuries. Even at this early date, the issue of metrics was raising its head. They immediately realized that different sized employee populations in the labor intense steel, iron, and other industries could quickly become skewed depending on the size of the organization or location being measured. Comparing a plant of 3000 employees to a plant of 500 employees was like comparing apples and oranges. Furthermore, because overtime was so prevalent, even using head count was not going to work.

As a result of these efforts, in 1920, the BLS developed a Bulletin 276, "Standardization of Industrial Accident Statistics." This bulletin was the forefather of the future American National Standards Institute (ANSI) Z16.1 standard and was the first attempt to develop a system to measure accident rates.

The bulletin essentially assigned days for types of accidents. As an example, there would be a 6000 lost workday charge for fatalities and similar types of "charges" for other types of injuries (this arcane assignment of days would stick with us until the late 1960s).

In 1926, the Secretary of Labor called for a revision of Bulletin 276. The request was for a "Sectional Committee" to develop a new standard on industrial accident classification. This in turn led to the formation of a coalition to address this standard. The coalition consisted of the International Association of Industrial Accident Boards and Commissions, the National Safety Council, and the National Council on Compensation Insurance. These three agreed to serve jointly as administrative sponsors of the project.

Although the revision of the bulletin helped, there was a move toward "standardizing" the recording methods. In 1937, this Sectional Committee gave birth to the American Standard Method of Compiling Industrial Injury Rates. The American Standards Association (the forerunner to ANSI) was the approval authority. The standard was published in 1937 for "trial use."

In 1939, the committee reformed to review comments on the "trial" and issue a new standard. This standard became ANSI Z16.1. The committee was comprised of all the notable safety professionals of the time including H.W. Heinrich who represented the Travelers Insurance Company.

I would like to pause at this time to point out that Heinrich who represented his employer, an insurance company, was typical of much of the early safety movement. If one goes back into the history of ASSE, NSC, and much of the safety movement of the early 20th century, the insurance industry played a prominent role in the development of societies, organizations, and standards. In a sense, they had the "A Team" of safety professionals. Sadly, over the course of time, the metrics used by insurance companies and safety professionals began to take different roads. I will address this in greater depth later, but there is no doubt the insurance industry played a large role in the profession many of us practice today.

The Z16 standard went through several iterations up into the 1960s and the birth of OSHA. While the standard had some language and concepts, such as the assigning of days to different types of injuries, it made one contribution that makes it noteworthy to this day. Somewhere along the way, changes in the standard million man-hours formula for calculating disabling injuries was conceived.

This formula is represented in the following manner:

Number of disabling injuries × 1,000,000 ÷ man-hours worked

Although it is lost to the mists of time who exactly came up with this formula, what we do know is that it became the prevalent method by which injuries and illnesses were measured for the next several decades (and counting). If the originator had a copyright on the formula and received royalties for every time it was used, they would be richer than Bill Gates. This formula has been used to calculate lost time accident rates, severity rates, recordable rates, and any other "rate" you could think of simply by adjusting the number of incidents or other numbers in the multiplier. It was not long before the "assigned days" became secondary to the "rate" that was generated by this formula.

Although the author of the formula is unknown, it is subtle genius. The million man-hours represent 1000 employees working an average year. Using the man-hours worked as a divisor allows for a leveling of the playing field across different size organizations. Consequently, if you have a plant with 1000 employees and another with 5000 employees, you can come up with a comparable rate between the two locations despite employee number differences. In other words, all things being equal, if the larger location has five times as many injuries, it is essentially equal to the smaller locations due to additional man-hours worked. It also addresses the problem of employees working overtime. Because it measures hours and not head count, it captures all the employee working more than the anticipated 40 hour week.

In today's economic lexicon, this would be called a *macrostatistic*. However, as time went on, the formula was used (or should I say abused) from the plant to the department level. As long as the man-hours were at a sufficiently high number, the formula was representative. However, as the man-hours would shrink, say to a department level, the formula was no longer viable. Any statistician would hardly give one a passing grade for using this formula without sufficient man-hours to justify the numbers.

This criticism was already being discussed among the safety profession when everything was changed (literally) by an act of Congress.

OSHA CHANGES THE FORMULA

On April 28, 1971, the world of safety changed dramatically. It was on this date that the Occupational Safety and Health Administration (OSHA) was born. Whether you curse the day or applaud it one thing was clear. A new visibility was given to the safety profession, which was to magnify and accelerate the "professionalization" of safety.

In order to get off to a running start, OSHA embraced many ANSI standards of the time for their 1910 and 1926 standards. Actually, this made a great deal of sense. There was no point in reinventing the wheel. The standards had been developed by safety professionals and were already in use in American industry. However, the one standard OSHA did not decide to adopt was ANSI Z16.1, the recordkeeping standard. Instead, they turned to BLS to set up new guidelines for "recordability." As if this were not earth shattering enough, they decided to do something that has caused angst which lingers with us to this day. They changed the formula for measuring incidents from one million man-hours as the divisor to 200,000 man-hours.

This change is reflected in the following:

Number of recordable incidents × 200,000 ÷ man-hours worked

There are varying theories on why OSHA and BLS decided to change the long used million man-hours formula to a denominator of 200,000 man-hours, but any discussion on the matter is a moot point at this juncture. As we are now aware, the concept was that the new formula would represent 100 people working a normal man-year, using 2000 hours as the average work year. Why the lesser representation of groupings of employees in the constant was selected (as opposed to 1000 people working a man-year) is anyone's guess. Although this change sounds like a nuance, the reverberations of that switch are still being felt to this day.

For those organizations with international operations, they now had to keep two sets of books. Although the United States switched to the 200,000 man-hour formula, the rest of the world had been using the million man-hours formula and saw no compelling reason to switch (nor do they to this day).

As fate would have it, I was one of those people with international operations. I soon found myself keeping one set of books for US operations and another for our international operations, all of whom were (and still are) using the million man-hours formula. This problem became exacerbated when I had to generate reports for upper management with an asterisk by the US numbers explaining the vast difference in rates. This is best explained by the following scenario.

If I had two locations, one in the United States and another in Spain that incurred 30 "recordable" injuries, I would end up with the following numbers using the two formulas:

$$US Rate$$
$$30 \times 200,000 \div 2,500,000 = 2.4$$

Now while using the old million man-hours formula with the identical losses, we see a different number:

$$Spain Rate$$
$$30 \times 1,000,000 \div 2,500,000 = 12$$

One can easily see why with the use of an abstract number like an incident rate confusion would reign supreme when the same number of incidents produced such dramatically different numbers.

As the world turned more and more into an international marketplace and the movement toward the Independent Business Unit (IBU) concept took hold that mixed domestic and international plants in one bag, the confusion on rates only got worse. Every report required a lengthy explanation of the two numbers, both of which were abstract.

Although we could argue the logic (or illogic) of OSHA's decision to change the long-used formula, the larger impact was on the perception of the numbers generated. The OSHA numbers being much lower make the losses seem much lower. The accusation of my international units was that we were artificially lowering the number to make it look better. Everyone started to lose sight of the number of incidents being the key factor not the formula to calculate the rate. But there is no denying that using the smaller multiple, the OSHA formula cuts the rates by one-fifth of the "international rate."

It is said that perception is 90% of reality. The perception here is that we were playing with the numbers to look better. If ever there was a case for finding a new metric that addresses incidents and loss, this was a compelling one.

THE ABUSE OF THE SYSTEM

Currently, there is a sense that lagging indicators are a bad thing. Far from it, lagging indicators are quite necessary. However, the bad reputation that lagging indicators in safety earned was from the abuse of the simple macrostatistic to which I earlier alluded.

We have to understand a simple concept of macro- and microstatistics. I like to keep things in simple terms, so let us define these two this way. A macrostatistic would be the one used at the corporate, division, IBU, or whatever upper level of your organization you would describe. In contrast, a microstatistic would go down to the department or small unit level where the man-hours are so small that a single incident can skew the numbers.

There is no "set" number that triggers macro and micro. However, if you have drilled down to such a level that it would take you all year to recover from one or two total recordable incident rate (TRIR) incidents, you are operating in a microatmosphere.

This abuse of a recordkeeping system that was built for macrostatistics is the first problem. The second problem is that it is designed to track one statistic: employee injuries and illnesses. A statistic that even when limited to that narrow parameter often does not make sense.

Let us examine this problem in greater depth.

3 The Body Count Dilemma

Using the employee injury rate as your sole measure of the success of your safety efforts is like trying to measure the volume of a silo by just calculating the height.

C. Gary Lopez

Although I do not consider myself a great philosopher, I will stand by the aforementioned quote. While considering metrics and measurements for the safety profession, perhaps no greater problem exists than what I have come to call "The Body Count Dilemma."

Admittedly, I receive many negative comments about what is construed as my cavalier treatment of employee injuries and illnesses by using this term. However, I use it not to take such injuries and illnesses lightly. Quite the contrary, I understand preventing employee injuries and illnesses is one of our primary objectives in what we do as a profession. Nevertheless, the operative word in the previous sentence is "one." I use this term to point out a serious shortcoming, or should I say limitation, in our thinking as a profession when we approach managing the risks of our respective operations.

I stole the term from the Vietnam War. Vietnam was a very different war for America. Unlike World War II where you could follow our progress toward victory by looking at a map of territory taken and front lines advancing, Vietnam had none of that. This presented our military with a conundrum. They had to demonstrate if we were winning or losing. Consequently, the military (or someone advising the military) decided on an odd method of measuring our progress in winning the war. Weekly they reported the "body count" of slain enemy soldiers.

As history proved, and as any military historian can now tell you regarding that conflict, this was not only a poor metric of measuring our progress in defeating the enemy but also a deceptive one.

In a sense, we have done the same thing with the "Body Count" metric we use and are suffering from some of the same illusions. Before we go too far down the "let's get rid of it" road, I want to point out to the reader that it still has value. More to the point, I am not endorsing getting rid of it. We just need to understand the limitations of the measurement and how to apply it wisely.

As discussed previously, this was a good macrostatistic for identifying injury/illness rates. Although it was not a bad metric for when it was developed, we need to ask ourselves if it is comprehensive enough to measure what the modern safety professional now does as their scope of work. Or perhaps worse, if it is limiting that scope.

Either intentionally or more likely unintentionally, we have made it the ultimate benchmark of success or failure in what we do. We have promoted it to upper management as the measuring stick of success. We have become so hypnotized with this number that it has driven us to put all other potential sources of measurement in the back seat. The impact to our profession has been crippling. Here is why it is a problem.

As safety professionals, we have moved into the realm of managing risk on a greater scope than we ever have before. Risk that goes beyond employee injuries and illnesses. Consequently, to declare that success, or failure, of our mission is measured in only employee injuries and illnesses ignores the risks we manage in other areas such as fleet, property, and so on. I would submit that this "tunnel vision" view of metrics is a symptom of 20th-century thinking that our profession has grown beyond. Unfortunately, one has only to look at 21st-century examples to see that this narrow view of our mission (and subsequent role as a profession) is alive and well.

FAILURE OF THE MANAGEMENT SYSTEMS APPROACH IN RECOGNIZING THE SCOPE OF RISK

As the 21st century dawned, the safety profession was moving full steam at developing "management systems" approaches to managing the risks our organizations face. These systems approaches were driven by the W. Edwards Deming approach toward quality that set off a movement in many industrial sectors of the country, especially in the automotive industry. The approach was based on a Plan, Do, Check, Act mantra that was turned into a set of ISO standards that laid out a path for this movement.

It was not long before, first, the environmental field and then the safety field were attempting to duplicate this "systems" approach in their respective fields. The result was the emergence of several safety management systems standards such as the International Labor Organization's MEOSH 2001 document and the British Standard Institute's OHSAS 18001 document as tools for our profession to use as guidance. However, I will focus on the two leading examples of these management system standards that emerged at the turn of the century. They were the ANSI Z10 Occupational Health and Safety Management Systems Standard and the more global ISO 45001 Occupational Health and Safety Management Systems Standard.

As an example of how deeply this "Body Count" thinking has permeated into our mentality as a profession, one has only to look at the scope/purpose of both of these documents (italics are mine):

ANSI Z10

1.2 Purpose

The primary purpose of this standard is to provide a management tool to reduce the risk of occupational injuries, illnesses, and fatalities.

ISO 45001

1 Scope

This document specifies requirements for an occupational health and safety (OH&S) management system, and gives guidance for its use, to enable organizations to provide safe and healthy workplaces, by preventing work-related <u>injury and ill health</u>, as well as by proactively improving its OH&S performance.

You will note that the goals of these two cutting edge documents written by some of the finest safety professionals in the country (in Z10's example) and in the world (in ISO 45001's example) are still mired in just viewing employee <u>injuries and illnesses</u> as our mission.

(Note: I would be remiss as an author not to point out that I sat on both committees. My pleas to expand the scope of ISO 45001 fell on deaf ears, but apparently I experienced some success with Z10 since they have expanded their scope in the latest version of the standard.)

I realize this is where the reader begins to question if I really "get it" in terms of what we do as a profession. I reiterate that I am neither questioning our need to address employee injuries and illnesses nor giving it a low priority. I am not discounting how important this part of our mission is for our profession. I understand that the mission we have as safety professionals is saving lives and preventing injuries and illnesses. However, that is not the total scope of our mission and we should not fall into the trap of thinking that other risks our organizations face are secondary in nature or worse, viewed as not being our job. As some recent disastrous accidents have demonstrated, ignoring these risks or assuming someone else is addressing them can have devastating results.

THE EXPANDED ROLE OF THE NEW SAFETY PROFESSIONAL

This is a book about metrics, not organizational structure or safety management philosophy. Nevertheless, it must be noted that because I started with the premise that "We manage what we measure," I can certainly not take the position that metrics have no impact on either.

The implications of restricting our role to just employee injuries and illnesses are staggering. If we are developing our management systems around only the concerns for employee injury and illnesses, we are missing our greater call of managing *total risk* for our employers. We are also in danger of bringing into question the worth of implementing these complex management systems. This is not to suggest we should not do so, but to stop at not considering the larger implications of what we do in managing risk is inexcusable.

Our need to measure risk in more than bodily injury has been evident to our insurance industry for some time. However, as safety professionals, we seem to lag in taking those numbers and devising metrics for use in organizational reporting and more importantly in our strategy toward measuring and controlling our risks.

There has been ample evidence in accident investigations of some large incidents that we still fail to grasp this concept. In testimony following the Texas City explosion, the company executives attempted to defend their safety program by pointing out their excellent employee injury statistics. In defense of BP their employee injury numbers were excellent and their safety program was not exactly nonexistent. However, attempting to apply this metric to an incident that clearly moved into another area of risk was greeted with great derision by one of the members of the the Baker Committee who flat out asked what such numbers had to do with the incident they were discussing. Clearly, even to a non-safety professional these numbers had nothing to do with the cause of the incident.

In some large incidents, significant damage can be done to an organization without an employee suffering a scratch. Perhaps, the greatest recent example of this is the tank farm fire in Houston, Texas on March 17, 2019. The fire that occurred on a tank farm owned by Intercontinental Terminals Company (ITC) was disastrous. The fire eventually consumed 11 tanks and resulted in losses that are yet to be tallied, but will probably reach the one billion dollar mark or more. This was a huge fire resulting in both environmental and property damage. That this loss was devastating to ITC goes without saying.

Of note, not a single employee was injured, nor were any of the emergency responders. Consequently, using a "Body Count" metric, one could speculate that no real damage was done. However, would any competent safety director want to approach the board of ITC, or any organization that suffered such a disaster, and take the position that it is a shame about the incident, but don't worry our safety record (judged on employee injuries) is still excellent after this event? As a safety director for that company, would I want to report to my upper management that the economic impact of this disaster was terrible but don't worry that our important metric, the lost time accident (LTA) rate, is intact? How long would I keep my job with that approach?

However, the true danger in not having a metric that would make this kind of loss inclusive is how it would impact our risk management efforts. If I or any of my management team are not being assessed on compliance with this type of risk, what does that do to my thinking? Would it be possible that I might even omit this hazard from my risk assessment process? Or would it influence any significant amount of capital requests to make improvements on that tank farm?

I will provide an example of this type of conundrum in a later chapter. It is a natural instinct to focus on the risks we measure. One would hope that the metrics alone would not impact our risk assessment decisions, but impacting our risk assessment priorities is another matter.

What if due to the lack of metrics addressing these priorities they were not properly elevated which in turn resulted in not putting in controls to address the risk? What would your excuse be if you had not approached putting controls in place, engineering or otherwise, to prevent this disaster with the same vigor for those controls that impacted employee injuries and illnesses? We cannot be hypnotized with our one metric to the detriment of other risks. We cannot let that metric blind us to other risks.

On another note, we tend to develop our roles around what we manage. If we are focused solely on employee injuries and illnesses, we can find ourselves viewing the other lines of risk such as property damage, general liability, and fleet, to name a few, as the purview of others. In many organizations today these areas of risk are viewed as the remit of seperated departments, by not only our fellow safety pro-fessionals but also upper management. They are considered the responsibility of the insurance department. In fact, it has led to a common schism in most organizations where the insurance department reports to the Chief Financial Officer (CFO) and the safety department reports to the operations management.

Compounding this dysfunctional approach, in many instances, the two depart-ments barely work in a symbiotic relationship other than to pass on loss data. This can lead to a dysfunctional relationship within the management team of an orga-nization as well as a breakdown in communications that results in poorly applying organization assets.

There is an old saying that we must be careful what we wish for we might get it. My wish was granted when I was put in charge of an international business unit's (IBU) safety, health, environmental, insurance, and security departments. How did that happen? The president of the business unit had heard me speak of the symbiotic relationship of these various areas and how it made no sense to have them scattered across the organization. It was simply inefficient to do so. Apparently, my argument was a compelling one. When he was appointed as the head of the IBU, he selected me to become his VP of EHS and Risk Management. My wish was granted.

I will speak more on this topic in a later chapter, but my first realization was how different (or completely lacking) the metrics were in each of these disciplines I now had under my purview. It also became clear to me how much of our planning and program implementation was driven more by regulatory compliance and whim than by a targeted process to address the risks of the organization. The sole metric we were reporting to upper management, and therefore was garnering any attention, was for employee injuries and illness rate.

This latter revelation should have come as no surprise. If you have been in charge of "employee safety" and have been pushing the LTA as "the number," then where are you going to focus your efforts? The effect can also trickle down. If you are measuring a plant manager or department supervisor's safety bonus on LTA rates then on what will they focus? The downstream effects quickly become evident.

As safety professionals, we need to think more holistically. In my new assign-ment, it was not long before I was faced with such a conflict of interest within my department. Several months into the job we suffered an incident at one of our warehouses.

WAREHOUSE INCIDENT: PART I

At one of our warehouses, a forklift driver had run into an in-rack sprinkler head. I will get into this incident in a later chapter in terms of hidden costs, but for now this incident illustrates the different thinking that can come from metric tunnel vision.

Like many incidents before this one, it started with a straightforward enough investigation, but unlike previous incidents, we were now operating under a new in-vestigation and value metric system I had installed to track the total cost of an incident.

My subordinates involved in this case were my Director of Safety and Health and my Risk Manager. Metric tunnel vision reared its head early. My Director of

Safety and Health was only concerned about the injury to the employee. After a great sigh of relief that it was not a recordable incident he zeroed in on the lack of licensing procedures and the need for more closely adhering to employee training of forklift drivers.

My Risk Manager (which is a misnomer if ever there was one) was concerned with damage to the products in the warehouse. He zeroed in on what the cost would be and how the deductible amounts for the damage would be handled.

The final player in this scenario was the Plant Manager. The Plant Manager was first (and predictably) concerned about how long his warehouse would be out of service for shipping, and second, if the injury going to be recordable (which would impact the safety portion of his bonus).

The investigation revealed that the forklift driver had indeed not been properly licensed to drive that particular forklift. As noted, his injury was not recordable. However, I did discover that it was the third time we had taken out an in-rack sprinkler head in this warehouse alone. As fate would have it, the previous two incidents also resulted in no employee injuries. Under our metric injury tracking system, this made both incidents nonserious and they did not impact our safety record. Because we were not using metrics such as tracking costs associated with these incidents, the investigations ended with employee discipline and training recommendations. In other words, they were invisible.

Under our new investigative process, we were switching over to using value-based metrics as our measure, in other words, tracking what they cost. Furthermore, we were digging deeply into not only the direct costs but also the hidden costs. In other words, we were tracking the total cost of the incident.

As this new investigation unfolded it forced us to ask why certain systems were in place, (the in-rack sprinklers), that contributed to the incident. This led to inquiring who had decided we needed in-rack sprinklers. Much to my surprise, I found that it was "my Risk Manager" who had instituted this requirement, not only in this warehouse but also in several others. When I asked on what basis this decision had been made, the answer was "because the insurance carrier asked for them."

I would like to point out that this is not a negative reflection on the Risk Manager. After all, he was simply a creature of his environment, and a creature of his background. Our hiring practices were not dissimilar to other organizations. When we needed someone to fill the insurance manager's job, we looked to the natural well of talent for this skillset: insurance companies. Consequently, like many in his position, as a former employee of an insurance carrier, he was well trained by them. He thought like an insurance company person and was sympathetic to insurance company approaches and recommendations to prevent loss. Which quite frankly was not all that bad. What was bad was that I had not set new operating parameters for him and had assumed he would understand that now he was operating on the "other side of the ditch" that his approach to risk controls was going to have to be viewed from a new perspective. That one was on me. We set forth to fix the mistakes in our system and apply metrics accordingly to future such incidents.

Our past sins were not finished with us yet. As events turned out, the post-accident investigation and recommendations (from the frontline supervisors) were focused around forklift driver training. Why? Because the safety department dutifully trained the supervisors how to investigate accidents and these investigations were focused on employee injuries and illnesses. Why would they focus on anything else?

In this case, the employee injury cost was negligible. The property damage, warehouse shutdown time, products damaged, and disposal costs, quite to the contrary, were anything but negligible. However, none of these would have been investigated under our "old" system. In fact, we did not even have a mechanism in place for tracking some of these costs (I had to get corporate clearance for the accounting department to release production cost figures for me to track some of the losses).

At this juncture, I would like to say it could not get worse, but we were just getting started.

As I dug deeper into who paid for the entire cost of the incident, I was stunned to discover that the "deductible cost" of this incident was handled by "corporate." The Plant Manager neither cared nor knew what the cost was for the incident. Which brought me full circle to having to find my Risk Manager blameless for not making an issue of this after the first two accidents. Why should he be blamed? No one was feeling the pain. Corporate was picking up the tab.

What this constituted was a clear case in which the metrics we were not tracking, the dollar cost of the incident, was aiding and abetting in "hiding" our losses. It certainly was of no consequence to the Plant Manager because it was not an operating expense for him, he was not billed for it and in essence the "good fairies" took care of the payment of the deductible. As noted, his major concern was shut-down time of the warehouse.

I learned some hard lessons from this incident. The first was that we lacked any kind of accurate metric to measure this type of incident. The downstream effect was that instead of focusing on our failed system controls (in-rack sprinklers), we were focused on system controls, such as training, that would be of dubious value. The larger issue, the cost of the loss, had become a "non-issue" because the money paying for the fix was invisible. Instead of reviewing our system for design specifications that were driven by an outside party (our insurance carrier), we were worried about forklift driver training.

This is a classic case of how the metric defines the role. Because we were in the traditional metric structure of the employee injury driving the process, we were missing the forest for the trees. Had we been in the traditional management structure of the Risk Manager and Safety Director reporting to different department heads, it would have been business as usual and we would still be installing the same fire protective system and suffering the same losses because of a "protective system" that was protecting us in one way and hurting us in another.

I will address this incident in greater detail in a later chapter, but let it be said that lessons were learned by all including why a symbiotic relationship is necessary between safety and insurance. The two are on the same mission, but not always approaching it the same way.

This was one example I was to run into when I saw the bigger picture. As a profession, we need to see the bigger picture.

REFERENCES

ANSI Z10 Occupational Health and Safety Management Systems 2005. American National Standards Institute.

ISO 45001 Occupational Health and Safety Management Systems 2018. International Standards Organization.

4 Using Data Properly: Avoiding Garbage In, Garbage Out

There are three kinds of lies: Lies, Damned Lies and Statistics.

Mark Twain

The quote from Mark Twain was his way of saying that even back in the 19th century when we did not have computers to amass statistics, that man was still using numbers as a method to support weak arguments, to rationalize what you wanted, or to blind the masses with irrelevant numbers to prove a point.

Fast forward to the 21st century and Mark Twain would not recognize our information-rich society. But he might recognize some of the same tactics.

Modern technology has provided our profession with more data than we could have dreamed of in our wildest imaginations just 30 years ago. The amount of information that is at our fingertips through the internet and social media is overwhelming. The amount of data we can gather because of technological advances in, what I call, the "Electronic Revolution" is equally overwhelming.

We are living in the era of the computer where information can be sliced, diced, and stored immediately. We can make incredible graphs and can create "dashboards" that can wow you with beautiful graphic representations of the data.

As if this were not enough, we are moving into an age where technology is giving us even more raw data to input into the system. As an example, it was not long ago that the measure of fleet safety was the accident rate per million miles driven. Now? We have sophisticated fleet safety data in the form of telematics that can tell us everything about a person's driving habits. We can track acceleration, stopping distance, and speed. We have dashcams and tracking devices that can watch our every move. We are rapidly reaching, or should I say we have already reached, what I call *information overload*.

Assuming all of the data is reliable, which is an assumption we should never automatically accept, several questions come to mind. What does it all mean? How do we present it so it makes sense? How do we present it so that we don't dilute the message we are trying to deliver? What does upper management consider the best measurements?

Recently, I did a consulting job for a company with a very large fleet. They are keen on using the telematics data to improve their fleet safety. However, once they had all the data, they were overwhelmed. They asked me which was meaningful and which was not. Now there was a good question. One to which I would not presume to know the right or wrong answer. One to which solid research probably needs to be applied to give a credible answer.

When I say I would not presume to know the right answer, what I mean is that I do not have enough information about how the telematics translate into safe driving to give a definitive answer or conclusion.

Do I have a sense of what is correct? Of course. Do I know that when people drive slower they have fewer accidents? Of course. Do I sense that if you are not braking hard, you are probably driving at a safer speed with proper distancing? Of course. However, are any of my conclusions verified by hard research or my opinion? The answer is the latter. They are my opinion, professional or otherwise, and I could be wrong without that verifiable research backing me up.

As an example, let us say we are tracking who is driving outside of the speed limit (new systems have what they call geo-fencing, which means they can identify the speed limit by GPS as part of the telematics data). If a driver is going 10 miles an hour over the speed limit on an interstate, do I treat that the same as 10 miles an hour over the speed limit through a school zone? Is speeding a better indicator of unsafe driving then, say, hard braking? Finally, what do I do with my errant driver? Coach them? Send them to more defensive driver training? Fire them? The answer to all of these questions is "it depends."

This represents the classic problem we all face with having a boatload of data but no reference points for using it correctly. Going back to the previous example, what if we decide to use speeding ten miles over the speed limit as a benchmark. Without the hard research I alluded to earlier, is this a solid indicator of behavior that impacts the safety of the operation of the vehicle. Or is it "intuitive" knowledge on our part. As professionals, we can only take "intuitive" to a certain length. Otherwise, in all honestly, we are guessing not providing solid answers. To use this metric properly, we must be able to correlate violating the norm in this area to increased accidents. Otherwise, what was the point?

We will deal with this issue in the chapter on leading indicators, but one of the great challenges is taking the mountain of data and using it to attain an objective of improvement. By "improvement" I mean reducing risk.

PARETO CHARTS, PIE CHARTS, HISTOGRAMS, RUN CHARTS, CONTROL CHARTS ... ET AL.

I have often been accused of not recognizing the value of charts, especially as manifested in some of the modern graphics we have at our fingertips. I plead guilty to this accusation. I am old fashioned about this. I am more concerned with the message we are trying to deliver with our data than how pretty the message is that we are delivering.

However (after a heavy sigh), I will admit that the tools at our disposal to illustrate our points are considerable. Gone are the days when only a histogram or pie chart was the sole method of a visual representation of the information you were presenting. The electronic revolution has given us many options and variations to present this data.

My warning is that this is a two-edged sword. I will give an example in a later chapter of how burying the recipient of the report we are generating in a blizzard of charts and graphs can actually blur the message instead of enhancing it.

At the end of the day, we have to remember that the message we are trying to deliver is the objective. If I can deliver it effectively by writing with crayons on a wall, I will use that as my delivery method.

In this book, I will neither go into the definition nor the logic behind each type of graph and chart, and also not on how they are compiled. Others have done an excellent job of this already and in great depth. I will remind the reader again that these are vehicles, which are representations of a point you are trying to get across regarding your metric. Admittedly, there are those that find it easier to comprehend such metrics in a graphic form. Who are they? Like many other issues dealing with metrics, the answer is "it depends" on your audience.

In later chapters, I will deal with balanced scorecards and working with management in selecting metrics. To a large degree, these selections will determine what graphics, if any, make sense and to whom.

Nevertheless, I would be remiss if I did not point out some of these various vehicles we have available to us to deliver our metrics message.

Below are illustrations of some of these graphics.

While I have turned over the explanation of how each of these charts was designed and how they are used to others, what I will address is the general concepts of when to use them with upper and middle management. I say "general" concepts, because, as I have just stated, there is no single fixed formula. We must determine what the appropriate metrics are for our organization, which is our primary goal. How they are presented is secondary to that goal and will differ at the various management levels. Let us start by dealing with upper management.

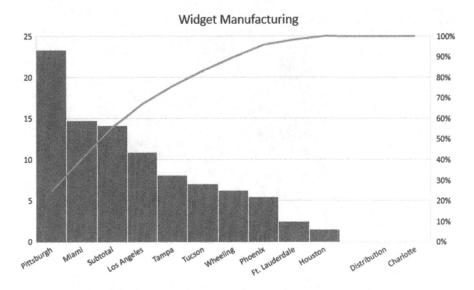

FIGURE 4.1 Histogram or Bar Chart

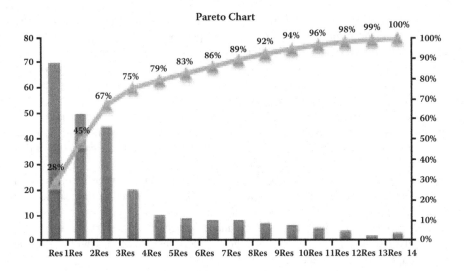

FIGURE 4.2 Pareto Chart

When dealing with upper management, they rarely want to know how the sausage is made detail by detail. They want the executive summary. Is the sausage selling well?

I am sure there are exceptions, but, for the most part, there is a correlation between the higher you go on the management ladder and the corresponding time they have for reports. Simply put, this means top management has less time for reading reports, which means the less they want of the intimate details. Attempting

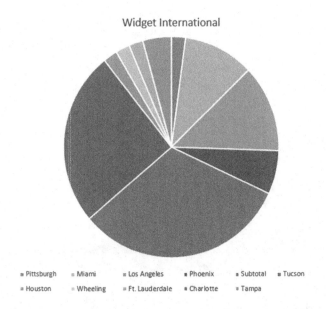

FIGURE 4.3 Run Chart

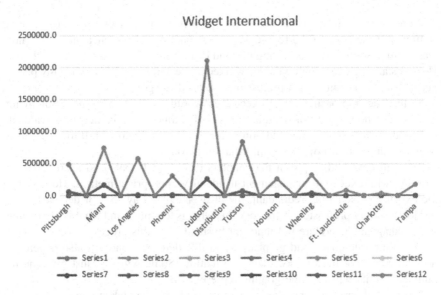

FIGURE 4.4 Linear Chart

to bury them with a variety of graphs and charts is usually fatal to delivering your message. While true that the "dashboard" concept has become very popular, this dashboard has to be set up to deliver your message succinctly along with your definition of success.

This is best illustrated by one of my "mistakes" in submitting one of my first reports to top management.

AN UPPER MANAGER'S VIEW OF THE WORLD

I once created an extremely thorough report that amplified the successes we were having in our IBU as a result of the implementation of our safety management systems in our operations. It was a work of art. A thing of beauty. Charts, graphs, and data – all beautifully illustrated. I submitted the report to the president of our IBU. It came back to me with a note. "Thank you for this exhaustive report and all of the hard work you put into it, however, I have a short attention span so in the future if you can keep it shorter it would be much appreciated."

While one might view this as a compliment, I knew this particular executive well enough to know that he was one of the smartest and most organized people that I had ever worked for. He was the definition of a leader. Decisive, a mentor to his subordinates, patient, and a man who created a team concept before it was popular to do so. There was nothing wrong with his attention span. You could have read him Tolstoy's *War and Peace* and not exceeded his attention span. This was his polite way of saying, "Your report is too darn long; if you can't say it on one page (preferably one paragraph), cut it down until you can."

None of my future reports exceeded one page. Any that did had an executive summary at the front of the report. My assessment of the true meaning of his note was verified sometime later during a round of golf in which I was paired with him. He was standing over a putt when he winked at me and said, "I like when my people get it. It's a rare trait." As a profession, we need to "get it."

Is this the approach that all executives would take? Perhaps not. But in my experience, the majority of them are pressed for time. Time is their most valuable commodity. Get to the point and don't attempt to wow them with fluff. This, then, begs the question of how much information is too much?

We will cover this topic in the chapters dealing with balanced scorecards and working with executive management to identify appropriate metrics. However, at the expense of sounding redundant, the answer will always be, "it depends." It is our job as safety professionals to work this what is too much and what is not enough with management. We are not innocent bystanders in this process.

As to the role charts and graphs play in this delivery, that will also depend on how an executive likes to see their information. Some will say, just give me the numbers. Some will love the graphic representation.

Where such graphics can play a significant role is when taken down to the more "tactical" level. As we expand on the metrics we deliver at the department and even plant level, these can be of great benefit. Instead of columns of numbers, the graphic illustration can deliver the point more clearly. This is especially the case when we post these metrics on plant bulletin boards for all to see.

There is also value in these types of charts if you are trying to make a "sale" for capital expenditure. They can be used to rationalize several data streams that support why you are requesting a particular capital expenditure that may be viewed as questionable. Supporting data can sway the attitudes of those reviewing the request if documented properly. When presented in a graphic representation, it can tell a story in a clear and sometimes compelling fashion.

IS ANY DATA BAD DATA?

Let us start with an assumption that the data you have is from a reliable source and is accurate. Can we conclude that all data is good data when applied to reduce risk?

One might conclude that any addressing of the "problem" automatically reduces risk. But to what end? Let us revisit our speeding example. If I am focused on a driver who is constantly exceeding the speed limit by 10 mph, does that tell the story? What if the excess speed is always on an interstate? What if the driver goes over the speed limit but maintains a good stopping distance between his vehicle and other vehicles on the interstate? Did I really contribute to reducing risk? More to the point, if our objective is to reduce accident costs, have I influenced those costs? What if the driver has a perfect driving record? Am I "coaching" a driver who is a nonproblem because of data that does not support my objective of cost reduction?

In some instances, the sorting through which data is useful and which data is indicative of the safety culture could be a fool's errand without the proper substantiation and correlation mentioned earlier. We must base our choice of metrics on hard facts, not gut feelings and urban myth. Or worse, simply because it is now available to us.

Even data based on some degree of truths can be distorted. Let us say we have data that demonstrates accident rates increase with speed. That is a fairly well-substantiated fact. However, is it 10 miles an hour that makes the difference? Or 20 miles per hour? We are back to the point, do we have the supporting research and knowledge to support the telematics decision.

What data we select and verify as accurate also has to be filtered. If you have two drivers delivering goods, hard braking in an urban driving route is going to be more common than hard braking on a rural route or interstate. Anyone who has driven in heavy city traffic can make this deduction. Consequently, what do you do with this data? Do you treat them equally? Finally, what do I do with a hard braker who has never had an accident in 25 years of driving? Pure raw data, like a raw ore, needs to be refined before we can use it.

THE DATA BARRAGE

Perhaps the worse at "data barrage" are insurance companies. They relish data gathering. In their defense, they have a long history of using data extensively in their underwriting departments. Furthermore, they do have a proven track record with much of their data. As a small example, their data can determine life expectancy (of people and things), high accident geographic areas, and experience with unprotected properties.

Their rationale is a sound one. The data they gather is extremely helpful to their underwriters in determining premium rates. It can also be helpful to insurers in calculating the risk of the particular lines of insurance. Since much of what insurers do is data-driven, it naturally begins to spill over into other areas of their operations.

One of these areas is, what they term "loss control." There is a plethora of accident-related data that insurers gather and make available not just to their loss control people but also to the clients they insure. You can get reports that tell you the part of the body injured, the time of the injury, how many years of experience the injured party had, the age group, the gender, and much more. You can get cost of injuries, cost of property damage, and costs broken down five ways from Sunday.

However, as discussed, all data is not useful data. Consider the following pivot table (Table 4.1) from an actual report that addresses workers' compensation cases in relation to date of hire.

While this might provide some fascinating information to a safety professional conducting research, what does it tell an executive?

None of the numbers are "normalized," so I can either conclude that I must get rid of all of my people after two years or after seven days. If I do not have an employee head count tied into this number, how can I determine the distribution of these losses.

The chart in Table 4.2 is an illustration of a similar problem. In this chart, I have broken out the locations by plant and reported their losses.

As with Table 4.1, Table 4.2 provides raw numbers of the losses, but what story does that tell? Plant A has losses that exceed Plant B's by 20 times the number. However, what if Plant A is working man-hours that exceed Plant B by 40 times?

TABLE 4.1
Worktime Spreadsheet

A. Hire _To_Acc_Date range: Displays Count & Avg for range of time passed from employee date of hire until reporting first incident

HIRE_TO_ACC_DATE_RANGE	Cnt	TTL.Inc	TTL.Pd	TTL.Outst	Avg.Inc	Max.Inc	%Cnt	%Inc	Avg.Lag	Avg.Age	Avg.Tenure
0 - 7 Days	8	$11,756	$11,756	$0	$1,469	$9,724	3%	1%	5.1	35.75	0.00
08 - 30 Days	21	$62,177	$62,177	$0	$2,961	$41,953	7%	4%	7.5	34.19	0.00
1 Month - 03 Months	58	$211,981	$205,825	$6,156	$3,655	$47,686	19%	13%	6.8	32.78	0.00
3 Months - 06 Months	46	$156,598	$140,332	$16,266	$3,404	$25,600	15%	9%	8.5	31.98	0.00
6 Months - One Year	48	$228,939	$189,619	$39,320	$4,770	$100,108	16%	14%	8.7	36.56	0.00
A Year - Two Years	45	$343,407	$316,101	$27,307	$7,631	$67,567	15%	21%	16.5	34.89	1.00
More than Two Years	75	$640,406	$453,614	$186,792	$8,539	$112,870	25%	39%	22.7	44.39	5.40
No_Data	4	$1,912	$1,912	$0	$478	$1,195	1%	0%	5.8	35.75	
Grand Total	305	$1,657,177	$1,381,336	$275,841	$5,433	$112,870	100%	100%	12.7	36.63	1.5

B. Tenure Range: Display Count & Avg for range of Length of Employment/Tenure on the date of the accident

Tenure Range	Cnt	TTL.Inc	TTL.Pd	TTL.Outst	Avg.Inc	Max.Inc	%Cnt	%Inc	Avg.Lag	Avg.Age	Avg.Tenure
No Hire Date	4	$1,912	$1,912	$0	$478	$1,195	1%	0%	5.8	35.75	
0 - 12Mths	181	$671,451	$609,709	$61,742	$3,710	_$100,108_	59%	41%	7.7	33.87	0.00
12 Mths - 24 Mths	45	$343,407	$316,101	$27,307	_$7,631_	$67,567	15%	21%	16.5	34.89	1.00
24 Mths - 36 Mths	23	$90,040	$54,014	$36,026	$3,915	$36,333	8%	5%	22.6	42.22	2.00
36 Mths - 48 Mths	15	$60,778	$60,778	$0	$4,052	$32,982	5%	4%	9.6	42.07	3.00
48 Mths - 60 Mths	7	$3,400	$3,400	$0	$486	$1,711	2%	0%	75.6	41.00	4.29
>5 Years	30	$486,187	$335,422	$150,766	_$16,206_	_$112,870_	10%	29%	17.0	48.00	9.47
Grand Total	305	$1,657,177	$1,381,336	$275,841	$5,433	$112,870	100%	100%	12.7	36.63	1.5

An even better question may be why Plant B, while working much less hours, has so many claims even though they are not expensive claims.

The age range tells us even less. Plant E is having most of its accidents in the age range of 25 to 34. Unless the age range of the employees at the plant is perfectly distributed, this information is virtually useless to a manager. What if the majority of employees at this location are in the 25–34 range and this represents a normal distribution of accidents for that range?

Even year-to-year comparisons become troublesome. Table 4.3 illustrates the losses for three years running.

Once again the charts are beautifully done, the information accurate but the value blurred. As it turns out, this company was in an acquisition mode over this period and nearly doubled its size. With that in mind, this would make 2015 a spectacular success. The losses were nearly $75,000 less than 2013 with twice the employee count.

However, as with the telematics data example, one must be cautious with these numbers. These numbers are based on the insurance reserve numbers for total incurred costs. Insurance companies will often set "reserves" for claims. Depending on the techniques used to set these reserves, the actual value of the loss suffered is sometimes not immediately evident. Most modern reserving techniques provide an initial reserve that is not fully "developed" or put more bluntly, accurate of the final cost. Normally, it takes as many as three years in time to see the final cost of some of the larger claims. Consequently, if I track the cost of an injury due to the initial reserve, I may find that I underestimated my real loss by as much as 30% (the rule of thumb used for calculating final developed cost). The numbers for the aforementioned incurred could look more like the those in Table 4.3.

TABLE 4.2
Multiplant Loss Spreadsheet

BusinessUnit	AGE_RANGE	Cnt	Ttl.Inc	TtL.Paid	TTL.Outst	Avg_Inc	Max_Inc	%_Cnt	%_Inc	Avg.Lag(d)	Avg_Age	Avg_tenure
Plant A	16 - 19	1	$301	$301	$0	$301	$301	0%	0%	2.0	19.00	1.00
	20 - 24	16	$146,665	$136,418	$10,247	$9,167	$47,686	5%	9%	5.8	22.19	0.38
	25 - 34	19	$92,709	$90,703	$2,006	$4,879	$42,695	6%	6%	3.3	29.47	1.00
	35 - 44	5	$29,407	$25,575	$3,832	$5,881	$23,374	2%	2%	5.2	41.80	1.20
	45 - 54	9	$50,197	$14,126	$36,072	$5,577	$36,333	3%	3%	6.1	48.78	0.67
	55 - 64	8	$127,307	$127,307	$0	$15,913	$48,707	3%	8%	22.8	58.25	0.75
Plant A Total		58	$446,586	$394,430	$52,156	$7,700	$48,707	19%	27%	7.2	35.31	0.76
Plant B	16 - 19	3	$17	$17	$0	$6	$17	1%	0%	2.7	16.67	0.00
	20 - 24	6	$13,227	$13,227	$0	$2,204	$12,953	2%	1%	4.2	22.00	0.17
	25 - 34	3	$101	$101	$0	$34	$101	1%	0%	1.3	28.33	0.33
	35 - 44	6	$6,484	$6,297	$188	$1,081	$6,326	2%	0%	2.3	37.50	0.33
	45 - 54	3	$801	$801	$0	$267	$801	1%	0%	5.0	51.33	0.00
	55 - 64	4	$1,743	$1,743	$0	$436	$1,726	1%	0%	5.0	57.50	1.25
Plant B Total		25	$22,372	$22,185	$188	$895	$12,953	8%	1%	3.4	35.04	0.36
Plant C	25 - 34	4	$1,001	$426	$575	$250	$575	1%	0%	12.5	32.25	2.00
	35 - 44	9	$18,261	$18,261	$0	$2,029	$11,492	3%	1%	21.8	40.33	3.00
	45 - 54	2	$0	$0	$0	$0	$0	1%	0%	78.5	46.00	6.50
	55 - 64	3	$16,845	$16,845	$0	$5,615	$16,845	1%	1%	32.7	58.33	3.33
Plant C Total		18	$36,107	$35,532	$575	$2,006	$16,845	6%	2%	27.8	42.17	3.22
Plant D	20 - 24	4	$8,717	$8,717	$0	$2,179	$5,730	1%	1%	22.3	22.50	0.00
	25 - 34	5	$5,282	$5,282	$0	$1,056	$3,071	2%	0%	10.2	31.40	0.00
	35 - 44	4	$3,511	$2,936	$575	$878	$2,216	1%	0%	4.3	38.75	0.50
	45 - 54	1	$11,403	$11,403	$0	$11,403	$11,403	0%	1%	45.0	48.00	0.00
	55 - 64	2	$107,791	$79,223	$28,569	$53,896	$100,108	1%	7%	121.0	58.50	0.00
Plant D Total		16	$136,704	$107,560	$29,144	$8,544	$100,108	5%	8%	27.8	35.44	0.13
Plant E	16 - 19	1	$0	$0	$0	$0	$0	0%	0%	6.0	19.00	0.00
	20 - 24	3	$3,446	$3,446	$0	$1,149	$3,159	1%	0%	15.3	23.33	0.00
	25 - 34	5	$33,672	$29,891	$3,780	$6,734	$18,435	2%	2%	4.2	27.00	0.60
	35 - 44	1	$63,458	$63,458	$0	$63,458	$63,458	0%	-4%	1.0	35.00	11.00
	45 - 54	2	$5,647	$5,647	$0	$2,823	$4,292	1%	0%	2.0	50.00	1.00
	55 - 64	2	$15,853	$15,853	$0	$7,927	$15,853	1%	1%	2.0	55.50	0.50
Plant E Total		14	$122,075	$118,295	$3,780	$8,720	$63,458	5%	7%	5.9	33.57	1.31
Plant F	20 - 24	2	$2,541	$2,541	$0	$1,271	$2,333	1%	0%	1.0	22.00	0.50
	25 - 34	6	$16,263	$9,838	$6,425	$2,711	$6,275	2%	1%	9.2	30.17	0.67
	35 - 44	3	$368	$368	$0	$123	$180	1%	0%	5.7	38.00	0.33
	45 - 54	2	$9,186	$1,894	$7,293	$4,593	$9,162	1%	1%	1.5	53.50	5.00
	65 over	1	$68,135	$64,405	$3,730	$68,135	$68,135	0%	4%	4.0	68.00	11.00
Plant F Total		14	$96,494	$79,046	$17,448	$6,892	$68,135	5%	6%	5.8	36.71	1.93

TABLE 4.3
Three-Year Total Spreadsheet

Cal_Yr	Cnt	Ttl.Inc	TtL.Paid	TTL.Outst	Avg_Inc	Max_Inc	%_Cnt	%_Inc	Avg.Lag(d)	Avg_Age	Avg_tenure
2013	51	$525,397	$461,658	$63,740	$10,302	$100,108	17%	32%	22.1	37.00	1.35
2014	84	$610,043	$598,495	$11,547	$7,262	$68,135	28%	37%	21.5	38.55	1.76
2015	76	$450,659	$259,117	$191,542	$5,930	$112,870	25%	27%	5.6	37.79	1.57
	211	$1,586,099	$1,319,269	$266,830	$7,517	$112,870	69%	96%	15.9	37.90	1.59

Elaborating on this concept, let us turn to the aforementioned Table 4.3. If the 2015 results are reporting in Q1 of 2016, this, in all probability, means that the numbers are as low as 30% for 2015 and as low as 15% for 2014. That means approximately another $135 K of "developed" losses are coming down the pipe for 2015. This leaves you with the option of factoring that in or changing your numbers on a constant basis to reflect the increasing value of claims as they are "adjusted" to the real number.

TABLE 4.4

Three-Year Summary Sheet

Cal_Yr	Cnt	Ttl.Inc
2013	51	$525,397
2014	84	$701,549
2015	76	$585,856
	211	$1,812,802

TABLE 4.5

Three-Year Summary Sheet: Unadjusted

Cal_Yr	Cnt	Ttl.Inc
2013	51	$525,397
2014	84	$610,043
2015	76	$450,659
	211	$1,586,099

As a note, this is not to condemn the insurance companies in their reserve tactics. Nor is it meant to condemn their using dollars as a metric to measure claims. In a later chapter, I will actually promote the concept of using dollars instead of abstract rates.

What it does point out is that the numbers have to be carefully put in perspective. They also need to be agreed upon by management. Let us look at an example using the earlier "development" case as a basis.

Table 4.5 shows the original numbers while Table 4.6 gives us the adjusted numbers factoring in "development" of the claims.

This same scenario occurs even if I decide to normalize my data. If I decide to adjust my cost per employee losses by 30% upward to reflect the full-anticipated development of the cost, the management team has to under-stand and agree to this number adjustment. Otherwise, it will be felt that you are dealing with "voodoo statistics." This addresses two issues with data gathering. First, the accurate reflection of the true cost, and second, the nor-malizing of the data so that executives can easily make sense of the dollar amounts.

So let us look at the normalizing of data.

TABLE 4.6
Three-Year Summary Sheet Adjusted

Cal_Yr	Cnt	Ttl.Inc
2013	51	$525,397
2014	84	$701,549
2015	76	$585,856
	211	$1,812,802

NORMALIZING DATA

When I received insurance carrier reports in my past, I was always frustrated with the "raw data" such as was illustrated in the charts showed earlier. I had no complaints about the data itself. I found it to be very accurate (in most cases). However, the raw data did not allow me to compare my high man-hour locations with my lower man-hour locations on an equal footing.

Without normalization of data, we have nothing but a mass of numbers that do not tell a clear story. I will not call such information useless, but its value is extremely limited. Even if you are using it for a single location, there has to be a method of addressing increase (or decrease) in employee numbers, growth through expansion, and so on.

A similar problem existed for acquisitions. If we had acquired new locations or ramped up operations in our existing locations, the data did not adjust for the increase. When I say "adjust," I mean that even when it reflects the increased costs because of an increase in expansion or increased workload, one can't immediately comprehend the reason for the change without insider knowledge of the cause. As an example, at one point I had a division that had increased in size by nearly 50%, yet their losses were the same as before the expansion. Taken as raw data, they were not improving. Normalized, they had improved by 50%.

Even chargebacks to my business divisions and plants were not possible with the raw data. Using my above example, why should I charge the division that reduced experience by 50% the same share as last year? Without having a normalized number, I am punishing the innocent.

Moving this logic process down the process chain, if I wanted to put together an executive summary for the president or CEO of the business, the numbers I provided could be deceiving unless they are "normalized" data.

Fortunately, with today's computerized world, all one has to do is simply tell the carrier the format in which you want the data reported and you will get it. However, it is still on the safety professional to decide what that format should be.

Let us explore how the normalizing of data can completely change the picture of how loss is occurring in an organization.

TABLE 4.7
Histogram Chart

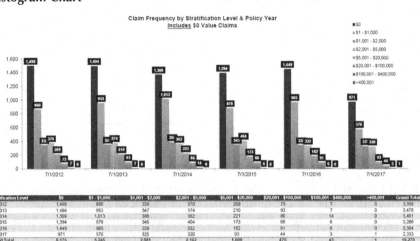

Stratification Level	$0	$1 - $1,000	$1,001 - $2,000	$2,001 - $5,000	$5,001 - $20,000	$20,001 - $100,000	$100,001 - $400,000	>400,001	Grand Total
7/1/2012	1,498	860	339	370	209	73	7	0	3,356
7/1/2013	1,494	953	347	374	210	93	7	0	3,478
7/1/2014	1,369	1,013	386	362	221	86	14	0	3,451
7/1/2015	1,394	878	345	404	173	88	6	0	3,288
7/1/2016	1,449	965	339	332	182	91	6	0	3,364
7/1/2017	971	576	325	320	93	44	3	1	2,333
Grand Total	8,175	5,245	2,081	2,162	1,088	475	43	1	19,270

Table 4.7 provides data from an actual report. This particular report deals with the stratification of claims.

As can be seen, the bar graphs (histograms) in Table 4.7 deliver a message of what the claims stratification is by use of colorful (or in this case, shaded) bars.

The first question we should ask is, what value is this metric to the reader of the report? Once again we are seeing six years of data with no indication if any conditions have changed in the workplace.

Even when presented in straight numbers format, we must ask ourselves what is the objective we are trying to accomplish with our numbers? Are we trying to find the areas of loss to attack or simply demonstrating where costs are going?

If we are trying to identify what departments are suffering the highest losses per size of the department, the raw numbers can be deceiving.

Table 4.8, for a fictional school district has had the data reported in a raw format. In this format, one can immediately conclude that the teachers are the problem and that resources need to be thrown at this job classification.

However, if we "normalize" this data and put it into agreed-upon "rates," we see another picture develop. Table 4.8 represents the kind of refined data that can tell a different story.

The information from the raw data in this table is normalized to allow for tracking several metrics. The one of most concern is the workers' compensation cost per man-hour. When converted to this new "rate" format, we find that it is not the teachers who are the largest loss problem. It is the bus drivers, by several orders of magnitude.

TABLE 4.8
Occupational Loss Chart

Top 30 Occupations by # of Claims	# of Claims	Total Incurred
Teacher	2931	$ 12,626,141
Maintenance	904	$ 5,619,368
Food Service Worker	882	$ 3,103,188
Teacher Assistant	754	$ 2,691,277
Bus Driver	717	$ 3,040,559
Custodian	600	$ 3,856,691
Para Professional	479	$ 1,685,124
BLANK / UNKNOWN	332	$ 1,363,400
ESE	307	$ 1,034,537
Substitute Teacher	256	$ 1,969,322
Administrative	248	$ 726,995
Bus Attendant	228	$ 762,754
Security	203	$ 1,433,813
Behavior Specialist	122	$ 280,420
Assistant Principal	117	$ 1,192,511
Counselor	79	$ 281,780
Speech and Language Pathologist	48	$ 71,067

TABLE 4.9
Normalization Chart

ABC School District
Loss Statistics
Q4-2019

Location	Report Rate	Comp Rate	OSHA Inc Rate	Report Number	Comp Number	Incident Number	LTA Rate	LTA #	Severity Rate	Workers Comp. Costs		Employee Number	Manhours
Occupations													
Teacher	2.9	2.6	2.1	72	65	53	0.9	22	0.0	$ 2.52	$ 12,600,000		5,000,000
Maintenance	6.2	6.2	4.0	31	31	20	2.4	12	0.0	$ 5.62	$ 5,619,000		1,000,000
Food Service	3.8	3.3	3.1	34	30	28	2.3	21	0.0	$ 1.72	$ 3,103,000		1,800,000
Bus Driver	17.3	14.7	13.3	26	22	20	0.0		0.0	$ 10.14	$ 3,040,559		300,000
Subtotal	4.0	3.7	3.0	163	148	121	1.4	55	0.0	$ 3.01	$ 24,362,559		8,100,000

DATA STREAMS

I always kept myself aware of the fact that data comes at you in two "streams" that are radically different. Both are of equal importance when addressing the risk to your organization.

There is always a temptation to be hypnotized by the obvious stream of data, which is the immediate losses you are experiencing. These "real losses" that we are faced with on our loss run reports create an overt look at the losses being suffered and the risks that are creating the losses. The result is a normal pressure one feels to do something about them. These are the classic losses most organizations suffer from back injuries, cumulative trauma injuries, slips, trips, falls, and other employee injuries that show up on loss control reports.

The natural inclination is to work on what preventative measures to put in place to eliminate or reduce the hazards causing these injuries. This is certainly worthy of our time, and I certainly do not endorse ignoring such losses and the actions we must take to eliminate, reduce, or control them. However, we must also be aware of the second stream of data.

I have already touched on the topic of our profession being cursed by the fact that our greatest successes go unrecognized. One of the greatest dangers a safety professional constantly faces is understanding the data that is not there. We cannot be seduced into thinking that the lack of statistical evidence of loss in certain areas translates into meaning there is no risk there. The reason there is no statistical data is often that the controls put in place to prevent these incidents are working. This "data vacuum" is a fact of life we must deal with on an ongoing basis.

As an example, when I was in the chemical industry, one of the greatest hazards we faced day to day in our operations, and that turned up consistently on our site risk assessments, was entry into confined spaces. Confined space entries were always a high risk for our risk assessment matrices. However, by putting in rigid controls, I can proudly state that during my time with my company, despite thousands of entries taking place across multiple locations, we did not have one confined space fatality.

As a safety professional, this is yet another limitation of the data we receive that we must recognize. There will always be pressure on the safety professional to address the losses directly impacting the organization. This is part of our job. However, you should not be a victim of "data hypnotism" that blinds you to other large hazards.

In my case, I was clear on the fact that we had no fatalities or any other injuries during confined space entries because of our almost fanatical obsession with good entry permits, use of air monitoring equipment, isolation of the spaces, and exceptional training of our people. The lack of loss data was because of these controls. Data cannot tell me what didn't happen.

An equally compelling argument can be made in the property risk area. Because a tank farm has not burned down yet does not mean that the risk of this happening is low. Data on immediate losses will not show this number. Nevertheless, another data stream one can use is what is happening in other

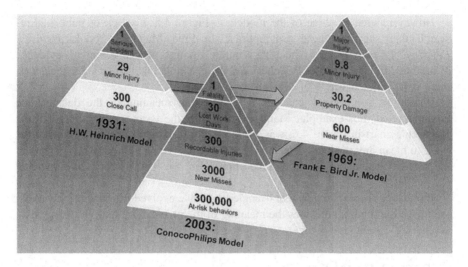

FIGURE 4.5 Accident Triangles

industries. Tank farm fires have resulted in huge losses to organizations. Use their loss experiences as a data stream.

In summary, never forget the "dataless" stream that is present at all times.

BLACK MAGIC DATA

Over the years, safety professionals have been "hypnotized" by what I have termed Black Magic data. This is data that is accepted as the "gospel" without anyone spending much time verifying the source and/or the accuracy of the data. Perhaps the most common Black Magic data source is the accident pyramids. Figures 4.5 and 4.6 are examples of these pyramids.

As can be noted in all three pyramidsin in Figure 4.5, there is an implied clear relationship between the number of near misses (or "close calls") using Heinrich's model occur in relation to fatalities.

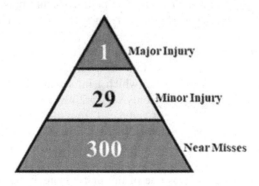

FIGURE 4.6 Heinrich Triangle

However, what all three of the models in Figure 4.5 is a lack of published research that supports these assumptions. In Fred Manuele's book, *Heinrich Revisited: Truisms or Myths*, he quite accurately points out that not only is evidence of hard research on much of this cherished data lacking, but also, in some instances, there is no proof it was ever done in the first place.

Worse, as terminology changes over time, the meaning of the data can be misinterpreted. For example, Manuele states that the term "major injury" in the Heinrich's chart in Figure 4.6, was explained by Heinrich in a later edition of his book to mean any medically treated injury, or what we would today classify under the OSHA system as a "recordable injury" in the Total Recordable Incident Rate (TRIR).

If one looks at similar accident triangles that have evolved from Heinrich's work, the top of the pyramid elevates the meaning of major accidents and, in many cases, is reserved for fatalities only.

THE FOUR-TO-ONE RATIO

Perhaps the second-most used Black Magic data is the 4:1 ratio so often quoted as the cost of accidents.

This 4:1 number was initially stipulated by Heinrich. There is also considerable use of this number as a 5:1 ratio. The theory of both ratios is the same. The "ratio" represents the hidden costs of accidents as opposed to the overt or "up-front" cost of accidents. I use the word "theory" because, once again, the validity of this data can be highly questioned.

If the safety profession truly desires to become a "profession," our data must be based on hard research and not questionable inherited tenets or, worse, myths. Some of which are virtually untraceable.

Similar to the above rate is another of the Black Magic data thrown at us is the "accident iceberg." Where this originated is unclear, but I will give Frank Bird credit since he was the first to explain it to me. To Bird's credit, he did the most work in terms of solid research to arrive at his numbers.

However, in a later conversation with Bird, he related that hidden costs varied so widely that attempting to place a simple single number on them as a multiplier was an exercise in futility. He remarked that the numbers could range from five times the cost to fifty times the cost. Hardly something we can standardize as a simple formula.

My personal research in this matter validated his latter statement. Accepting "theory" without validating can be fatal to your career. I once made a safety presentation to our plant managers in which I informed them that we had to re-evaluate our metrics since they were only seeing 20% of the true cost of incidents. Shortly after my presentation, I received a call from the president of the division, which I had delivered the presentation. His request was simple. Prove it! If I was stating the actual costs were a five-time multiple hidden costs, show him the evidence.

As a note to the reader, it was one of my first lessons in making sure that you have your facts right before opening your mouth.

My first move was to simply track down the source of this undisputable information regarding ratio's of loss. I mean it was gospel. Surely the source data was at someone's fingertips. However, I soon found that despite these numbers having been thrown around for years I could not dtermine the so-called "source" of this information, it was like chasing a ghost. No one could tell me where it came from. Urban myth in our profession is not new. We simply pass on unverified information and soon it becomes gospel. For years, contact lenses were was considered a danger in hazardous locations until we found through research they can actually be helpful.

At any rate, this set me (and my entire department) on a research mission of looking at every accident over the past year and every major accident over the past five years. We defined major accidents as any that had exceeded $10 K in workers' compensation reserves, fleet costs, or property damage.

My experiences in tracking what I came to term the "total cost" of an incident had some interesting revelations. To Bird's point, there were just too many variables in different incidents to simply slap a single number on them as the total cost of the incident. Even the "variables" varied from incident to incident. For example, I realized that the cost of production downtime was something we rarely measured. So much so that even our accounting people had to scramble to calculate this number in areas such as warehouse run time.

How did my study turn out? Interestingly enough, in the smaller incidents, the five-times figure was high. However, in the larger incidents, the five-times figure was too low by orders of magnitude. From a personal standpoint, I got lucky (I would love to call it skill and daring, but in the interest of veracity, I will call it what it was). The numbers in this case supported my assertion and actually demonstrated that I had understated the hidden costs.

Fate was on my side. I did not have to eat my words, or damage my credibility. Nevertheless, it taught me a valuable lesson. Before you accept all data as correct, check where it came from and the reliability of the data.

NATIONAL DATA

In Dan Petersen's book on measuring safety performance, he makes an interesting observation. The "integrity" of national statistics is questionable at best. I could not agree more.

Petersen used the word "integrity" a bit differently than I would. My choice of words would be "validity" of the statistics. Petersen considers them valid because there is a statistically valid database of numbers. To some degree, this is true. To another degree, it is not.

In my world, "valid" means they are accurate. Accuracy would include comprehensive in terms of what the numbers state. In this area alone, past numbers are problematic to some extent. As an example, the large statistical database to which Petersent refers to is the BLS database. Where do they get their numbers? Until recently, those numbers were mostly gathered from OSHA 300 form surveys. How confident was even OSHA in these figures? When OSHA pushed for their electronic recordkeeping standard, one of their main rationales was that they simply did not have a good way of accumulating accident data. While the BLS information was helpful and gave them a "sampling" of what various industry losses looked like, they were far from comprehensive. In this sense, OSHA was correct. The accuracy of the database was not a problem. The comprehensiveness of the database was the problem. OSHA felt they simply did not have data that was solid enough to determine where they should be prioritizing their resources in the future.

While the electronic recordkeeping rule is one of the most controversial that OSHA has enacted (and this is not a defense of that rule), their argument for the need for a more comprehensive data is hard to deny.

The BLS data is normally gathered in the traditional numbers of incident, lost-time accident, and restricted duty rates. While this information is of value for just looking at employee injuries and illnesses, it is hardly the last word on the success of a safety program. It also lacks the metric that makes the most sense, the dollar sign.

The National Safety Council attempts to rectify this by capturing these numbers in their annual publication *Injury Facts*. This particular publication breaks down accident costs in different areas, so to keep it in the same scope of other injuries we are dealing with I will concentrate on their section dealing with just occupational injuries.

In its 2018 edition of *Injury Facts,* the dollar figure of work-related injury losses is $170,800,000,000 in worker injury costs. However, once again we are dealing with the reliability of the data. Just how reliable is this number? In their footnote to this number, they state:

Source: Deaths reflect National Safety Council (NSC) analysis of data from the Bureau of Labor (BLS) Census of Fatal Occupational Injuries (CFOI). All other figures are NSC *estimates* based on data from BLS.

The italics on the work "estimates" is mine. Put another way, this says, "These numbers aren't completely accurate, but they are within an order of magnitude of accuracy." I suppose, on a national level, we have some wiggle room for estimating losses, but the bigger question becomes of what value is this to me when I bring it down to an organizational level?

The practical value of such numbers is to point out the potential for loss. Such numbers can be used to demonstrate exposures that are potentially present for an organization even if they have not suffered such incidents. Otherwise, we are just dealing with "Wow" tactics.

For using national numbers to rationalize capital expenditures and enhancements of a safety program, perhaps the most easily understood are those generated by the National Council of Compensation Insurances (NCCI). Why the NCCI? The NCCI probably offers some of the most reliable injury costs. Costs measured in dollars and cents. This is a metric we can all understand – dollar-based numbers. They collect these numbers from the insurance carriers for insurance modification ratings and risk categorization. These numbers can be of great value when used to support capital expenditures.

For example, their 2016–2017 numbers show that workers' compensation losses for fleet incidents averaged $78,293 per worker, while slips/trips/falls by workers were an average of $46,592 per worker.

Once again, these are national average numbers. Nevertheless, they can be used as "national average" numbers to make a point. Consequently, if I were trying to sell an upgrade of my fleet to include more safety features on the organization's vehicles, this number would come in handy as an ROI number. If I were attempting to sell an upgrade in flooring for a particular operation, these numbers once again would provide justification toward an return on investiment (ROI).

In summary, the lack of data for the modern safety professional will not be the problem. How to sort it, use it, and present it will be the key to success or failure in presenting results.

REFERENCES

Fred A Manuele. *Heinrich Revisited: Truisms and Myths.*
Dan Petersen. *Measurement of Safety Performance.* American Society of Safety Engineers.
National Safety Council. *Injury Facts 2018 Edition.*

5 Leading Indicators: What Are They, and How to Use Them

If you cannot put a number on it, it is not a metric. If you cannot normalize it, it is not a good metric.

Gary Lopez Law of Leading Indicators

At the expense of once again exceeding my "modesty scale" by quoting myself, my experience has shown that the aforementioned quote is a truism. How did I come to know this truth? Trial and error. Watching some leading metrics fail. Watching others succeed. In summary, I learned by experience.

As I noted in the Preface of this book, experience is a funny thing. What worked well for me, might not work for another in a different set of circumstances. It is why I continue to deliver the message in these pages that the metrics will vary depending on the organization. There is no simple "one-stop shopping" answer. Especially with leading indicators.

Beyond a doubt, the holy grail of safety metrics over the past three decades has been the quest for good leading indicators and putting them to good use. Leading indicators can be an invaluable tool for the safety professional and an organization's management team.

Many years ago, I began to experiment with leading indicators. As noted earlier, this is not a new approach to safety metrics; however, it has been a difficult one to implement and translate into an accepted metric. A metric that management at all levels will agree makes sense.

Despite the fact that this type of metric does makes sense, the difficulty in having leading indicators catch on as a way of measuring success has clearly shown that they have their limitations. I will attempt to cover the positives and negatives of using leading indicators in this chapter.

THE ROLE OF LEADING INDICATORS IN DEMONSTRATING MANAGEMENT SUPPORT

Early in my career, I had a boss who wanted me to accompany his boss into the executive suite to see what it was like to deal with the CEO of our company. He felt that my exposure to such rarified air would be good for me. He had no idea how correct his assumption would be in leaving a lasting impression on me.

My boss arranged for me to accompany his boss, who was the VP of Safety & Health, to this historic (for me) meeting. Naturally, at my level, I served no more purpose than as a warm-up for the meeting. The VP wanted to show the CEO one of our new "safety-trained" graduates they were hiring. Beyond that, I was there to listen. Which I did.

The purpose of the meeting was so that the VP of Safety & Health could obtain the support of the CEO for the new safety program he wanted to "sell" to our various plant managers.

As the meeting progressed, he made his salient points of why we needed to update the existing safety program and finally got around to asking the CEO for his support of the company's "new" safety program. The CEO who had been silent to this point nodded and replied "You have it, what do you want me to do?"

I will never know if our VP was that unprepared with a response to what I now know was going to be the obvious question or was he dumbfounded that the CEO acquiesced so easily, but he was literally speechless. Finally, he snapped out of his stupor and found the power of speech. He asked if the CEO would sign a new company safety policy. The CEO arched an eyebrow and replied that he would, but that he doubted a single piece of paper would represent his support. He was correct. But not in the way he had intended.

What he was asking of our VP was for the "actions" that were expected of him to represent his support. A corporate safety policy is wonderful, but like so many other policies and procedures, it is the management actions behind those policies and procedures that speak louder than the written word.

The CEO knew this. He had been listening silently to the sales pitch (as he had probably done a hundred times before from other departments), waiting patiently for "the hook," as I call it. What was it going to cost him? Perhaps this cost was not all financial in nature, but what we expected of him from a performance and extra-effort standpoint. What was he being asked to do?

In a sense, he was asking for the safety business plan to implement this change in culture and the role he was expected to play in implementing that plan. What this in turn meant from us was that he was asking for the leading indicators that he could support and put in place that could be drilled down to the frontline supervisors. Leading indicators by which the frontline supervisors, department supervisors, and plant managers could be evaluated. In other words, the actions that were expected of them.

Make no mistake, leading indicators are the "actions" we are asking of our supervisors and our employees that represent the controls and methods of reducing the risk to an organization. These actions should reflect the safety business plan that we annually develop to act as our path to enhancing and supporting the safety culture of an organization.

THE GOOD, THE BAD, THE UGLY

Perhaps the old Clint Eastwood Western is a good title for what we deal with in leading indicators. As I stated at the beginning of the chapter, leading indicators are the holy grail of safety metrics that our profession has grappled with over the past three decades. In many instances, it has been us, the safety professionals, that have had the hardest time developing systems to use them, making sense of them, and embracing them as a valid metric.

Of late, there have been several safety movements that are built around what we are doing right in our safety programs versus what we are doing wrong. From a metrics standpoint, this puts more critique on our traditional methods of measuring loss versus what we are doing right. The concept is to focus on the positive and not the negative of the results of our work.

While I applaud this new concept, the fact remains – at the end of the day, if I am measuring what I did right, I have to have a metric. In order to measure these "positive things," I will have to identify the "things" I am measuring. In other words, as I pointed out earlier, the actions expected to be carried out. Simply put, this means the leading indicators. Why is this paramount in supporting these new "positive" approaches to our safety programs?

You will not be able to walk into the plant manager's office at the end of the month and simply state you had a wonderful month filled with lots of positive things that happened. If they are a savvy plant manager, they will want some type of proof. We call that proof metrics.

I have no objection to these movements. In fact, they are based off of the positive reinforcement theories that have been used in the past to impact behavior. More to the point, they open the door for a measurement approach that emphasizes the use of leading indicators. However, there are limitations to how these leading indicators can be used effectively.

Perhaps the biggest myth, or misunderstanding, about leading indicators is that they are "predictive" in nature. Leading indicators do not predict where losses will occur. They are neither a failure mode analysis nor do they provide you with a crystal ball. However, they can be predictive to the extent that if the proper leading indicators are selected and applied correctly, they should reduce risk and, consequently, loss. Therefore, from that standpoint, they can be "predictive." However, this predictability is only as good as the selection of the correct measures that directly, or indirectly for that matter, impact loss.

Leading indicators, if properly selected, can have a significant impact on a safety program. However, here is where the "ugly" emerges in the process. One of the weaknesses of past approaches has been this very selection process. In too many instances, many of the selected indicators are a shotgun approach. For a variety of reasons (including perhaps the lack of a safety plan) in too many instances, little thought is given to what role the indicator is playing in the safety business plan in terms of impacting the safety program. This shotgun approach theory is that if it involves anything having to do with safety, it has to be a positive impact. Correct? Perhaps, but not when you are tying the leading indicator into specific metric measurements, especially if they are part of a larger plan or intended to impact key performance indicators (KPIs).

If, for example, you are looking at a strategic metric of trying to reduce workers' compensation cost for cumulative trauma injuries by 10%, you should have components built into your safety plan targeting this risk. Following this logic path, we would then want leading indicators tracking our success at implementing the interventions or controls you are putting in place to impact this risk.

This logic path for implementing leading indicators requires us to think first of the strategic level (which, in many instances, could be a KPI) and the tactical level, which would mean the front line where the interventions or controls are being implemented.

Clearly, this requires a strategy of selecting our leading indicators carefully and not the "shotgun" approach of selecting the classic leading indicators and hope for the best.

What are some of the classic leading indicators that have been subject to this shotgun approach. Although not all inclusive, the list below provides some of the "leading suspects" of historically used leading indicators:

- Training
- Inspection/Testing
- Audits
- Timeliness of accident investigations
- Near-miss reporting
- JSAs completed
- Safety meetings
- Attendance at safety meetings
- Employee observations
- Number of safety suggestions submitted
- Employee safety suggestions
- Procedure development
- Plant walkarounds

As stated, this is certainly not an exhaustive list, but all are action-oriented items that can be measured as a leading indicator. Before we continue, I will repeat the quote that led off this chapter. If you cannot put a number to these leading indicators, they are going to be impossible to measure. Further to this point, if you cannot normalize that number you have probably created a metric that can be as confusing as it is productive.

With that hopefully settled, let us move on with the assumption that you can put a number to these measures and normalize them (methods I will demonstrate in the following pages); we have to ask ourselves what the ultimate objective of these measures are in terms of impacting the safety program. Let us start with the most popular of leading indicators – training.

Training covers a lot of turf. I will not dispute the fact that all training has some positive effects on safety, but clearly some have more of an effect than others. The quality of that training will impact the end result as well. Especially when it is coupled with management backing.

Let us start with an example of an organization that has suffered some serious chemical handling incidents and wants to reduce or eliminate this type of incident. There are several ways this can be accomplished (all of which should be outlined in the safety business plan). One of the obvious will be training employees handling these materials. As with any organization, you only have "x" amount of time and resources to accomplish any task. As the safety professional, you are now put in the position of determining the true source of the problem. Why is this important? If you already had your people trained and the incidents continued, the problem may not be one of employee knowledge but instead a management issue. If you determine it is an employee knowledge issue and you are truly wishing to impact this particular item, you will realize training employees may have to take on a different tone than it did in the past.

Our goal is to impact the risk, not see how many leading indicators we can accumulate. If you are dealing with a management problem (they are not enforcing your internal procedures), your re-training will most likely be an exercise in futility. Your leading indicator would be better served as one that tracks management's adherence to procedure. Conversely, if you do determine that it is an employee knowledge matter, then training needs to be elevated in quality and the supervision must not only be made aware of the problem but also be assigned a leading indicator that measures if their people received such "new" training.

As an added note, you will see that in the aforementioned suggestion of this leading indicator, I did not suggest that the supervisor actually had to conduct the training. They may not have the expertise or training skill sets to deliver the quality training you are looking to implement. However, it would be fair to evaluate these supervisors to ensure that their people received the training. This avoids another of the "ugly" parts of leading indicators, which is assigning an activity to a manager over which they have no control or ability to achieve.

SELECTING GOOD LEADING INDICATORS

Let us start with a basic premise that we are selecting our leading indicators to impact KPIs that we have identified as suggesting success in our efforts. Before we charge down the road of taking the list of classic leading indicators and throwing them against the wall to see what will stick, we should first ask ourselves some of the following questions:

- Does it impact the success of my safety business plan?
- Has management endorsed this approach?
- Can I put a measure to it?
- Is the measure a number that can be normalized?

Then, as the system of leading indicators is implemented, we must keep asking ourselves follow-up questions:

- Is the leading indicator working as desired?
- Does it play a part in engaging the management team in meaningful activities of the safety business plan?
- Is management accepting the metric as valid?

As I stated earlier, perhaps of all of these points, #1 on the list is the most important and least thought out. In our rush to enact leading indicators, we sometimes just charge ahead and put one in place that historically we were told or believed to be a good one.

In case I haven't made my point, I would like to reiterate that when selecting a leading indicator, the first thought we should have is *"How will this impact the risks*

I am trying to eliminate, reduce or control?" Just randomly selecting any leading indicator will not do. There must be a correlation to what risk you are trying to eliminate, reduce, or control and the leading indicators you have selected to impact this risk. Leading indicators must be tied into this logic path to be truly successful. Is there a "halo effect" with almost any indicator? Probably, but this is about an organized approach with focus on our mission.

The second point is also important. Management has to be signed on to this effort. I will delve into this more completely in a later chapter, but they should have endorsed the safety business plan and, as tenet #7 suggests, buy into your metric to measure it.

Numbers 3 and 4 align with this concept. To be effective, leading indicators must be measured. Even if we are agreed on an activity, we must be able to measure it. Further to this is that the measurement must be normalized so that you have a level playing field.

Tenet #5 falls into the follow-up system where we must have to identify the simple question of if what we did is working. This normally means tying a lagging indicator into measuring the leading indicator.

Tenet #5 is a little-understood (or purposely forgotten) tenet that must be attached to every leading indicator. Whether the measurement is financial or an improvement is an established rate, there must be a corresponding measurement that validates success. This confirmation is normally done with a lagging indicator that demonstrates there is improvement in whatever goal you set. If the KPI was to reduce workers' comp by 10%, you have the lagging indicator in front of you that confirms success.

ALIGNING EXPECTATIONS WITH MANAGEMENT

To elaborate on these points, let us use a fleet example. Let us say that your organization is suffering significant losses from your sales fleet. Many of these losses deal with poor driving habits from your sales people, especially rear-ending of other vehicles. You, as the safety director, declare that a good leading indicator will be for all salespeople with company vehicles to go through a defensive driving program. You convince the CEO of the wisdom of this move. The CEO in turn instructs all his sales management teams to ensure their people are trained in defensive driving techniques.

As part of your new safety business plan, you set up a leading indicator metric to track completion of defensive training by your sales people. Your method of measuring this metric is based on a percentage of drivers trained by each manager. In the first year, you achieve 100% compliance of drivers completing the training. You have hit a perfect score on your leading indicator. You are so pleased that you make it an annual training requirement in your safety business plans that all sales people who drive company vehicles must go through defensive driver training on an annual basis.

Your leading indicator goal has been completely met by achieving training of all of your drivers. However, five years into this effort, you receive a call from the CEO. They are concerned that after spending thousands of dollars on training, not to mention the indirect cost of sales time lost in the field while training drivers, the accident costs for the sales fleet continue to climb.

Naturally, this begs the question whether or not the leading indicator chosen was the correct one. It puts to question if the accidents were a training problem in the first place. However, it also points to the fact that the CEO and the Safety Department had two different views on what metric measured success. While the CEO was measuring success in terms of accident costs for the fleet, the safety department was measuring success in terms of meeting the leading indicators metric of percentage of people trained.

I will address this matter of management and safety identifying goals that measure success in a later chapter. However, the point made in this scenario is that every leading indicator isn't necessarily a good one. In fact, some of the traditional leading indicators mentioned earlier don't necessarily impact traditional metrics such as lost time accident rates or dollar losses on lines of insurance. This is not to say that these activities might not yield other benefits, but being used as a leading indicator that directly correlates to the bottom line may not be one of them.

The lesson learned is that when developing any new metrics, management must be engaged in the process of not just agreeing they are good ones, but that their performance will be measured by them. Lesson #2 is that we must not lose focus on whether the leading indicator makes sense regarding what we are trying to accomplish.

USING LEADING INDICATORS EFFECTIVELY

Perhaps the greatest use of leading indicators is done at, what I classify as, the "tactical level" of an organization. In my life, in manufacturing, we would term this as the plant/department level of our organization.

Almost all modern organizations include "safety performance" as part of the merit review process for their management team. Whether you are the business unit president, plant manager, department manager, or floor supervisor, the merit review process to measure annual performance will no doubt include a safety portion. Normally, this is 20–25% of salary increase or bonus structure.

Note: If your organization does not have this process, the first thing you, as the safety professional, must do is have it instituted into the merit review system at all of these levels.

Using traditional lagging indicators to serve as metrics for this type of determination can be misleading and, worse, have disastrous consequences. As discussed in an earlier chapter, lost time accident rates, incident rates, and other such rates based on macro numbers do not work well when drilled down to the "department" level. These rates suffer from a variety of shortcomings, the worst of which is that when used in instances where there are not sufficient man-hours to create a "critical mass," they are more than misleading.

These rates suffer from other sins as well. First, they do not give the supervising manager a sense of control. Good managers want to know that they have control of their destiny. Whether it is production rates, quality, or waste, they like to know there are "things" they can do to impact these outcomes. While it is true they can do things to indirectly impact the LTA and incident rate, they feel better knowing they are being rated on if they did these things (and did them correctly) as opposed to an accident rate that is an indirect outcome of what they were asked to do.

Perhaps the greatest flaw in using traditional lagging indicators is the inability to recover from what could be a bad start. I am not supposed to use the word "luck" as a safety professional but anyone who has spent much time investigating accidents can tell you that in many instances, the difference between a minor incident and a severe one is luck. I realize that, in many situations, we make our luck, but clearly there is a compelling case that many incidents could have gone either way by the fraction of an inch, or time. Managers who feel they are a victim of this "bad luck" can get a sense of no hope of recovery in a worst-case scenario. This is easily demonstrated in the following example.

Let us say that a plant has set an annual goal of no more than .5 as an LTA rate and 1.5 as a recordable incident rate. As the ABC Company Plant Manager, I inform all of my department supervisors that this goal must be met by each of them or they will not pass the safety portion of their merit review – a portion that impacts their annual bonus. Why would I, the plant manager, do this? Because my bonus is being determined on this rate by my boss. Consequently, using good sound management techniques I have been taught, I hold my people responsible to the same goals I am being held. However, despite this sound management logic, there is a deep flaw in this thinking because of the chosen metric.

As a department supervisor, I now find myself being measured for my safety performance using the historical LTA rate and OSHA incident rate (TRIR). Nothing else. Just those two metrics.

As a supervisor, I have responsibility for a department of 50 employees. This means that the normal total work hours for my department (excluding overtime) will be in the range of 100,000.

Our year for safety performance is measured on an annual basis starting on January 1. In January, one of my employees who, despite wearing his safety glasses, reports that there is something wrong with his eye. This employee is not certain where it happened but since he noticed it at work, he assumes it was in the workplace. The employee is sent to the doctor for treatment. The doctor flushes the eye and prescribes a prescription antibiotic as a precautionary measure to avoid infection to the eye. That makes this injury "recordable," resulting in the following incident rate:

$$1 \times 200,000 \div 8333 = 24.0$$

Of note is the fact that I only have 8333 man-hours to use as my denominator since it is the first month of the year. This number represents one-twelfth of the man-hours my entire department will work over the year. Since a monthly safety report is generated, my department now clearly looks out of control.

In the following month, things get worse. I have an employee who reports that he has a hernia. He is not sure where it happened but feels certain that it is work-related. Now I have two recordable incidents one of which is a lost time accident. This results in the following number:

$$2 \times 200,000 \div 16,667 = 24.0 \text{ for an incident rate}$$

$$1 \times 200,000 \div 16,667 = 12.0 \text{ for an LTA rate}$$

Since it is an incident in the second month, I can use that month's man-hours for my monthly safety report, but it does not improve my numbers.

While these numbers make my department look horrendous enough early in the year, I have a bigger problem. If we extrapolate to the end of the year with no more recordable or lost time accidents occurring in my department, I still cannot meet my annual goal. The following math demonstrates the futility of my safety performance:

$$2 \times 200,000 \div 100,000 = 4.0 \text{ for an incident rate}$$

$$1 \times 200,000 \div 100,000 = 2.0 \text{ for an LTA rate}$$

As the department supervisor, I realize that, no matter what I do, I cannot meet the plant's annual goals for LTA of .5 and incident rate of 1.5. Consequently, this portion of my bonus is lost.

I can plea that I had no direct control over either incident, but this is irrelevant. That I followed all of the procedures and safety requirements that were asked of me is also irrelevant. That I am safety conscious and look out for my employees is irrelevant in this measure (but no doubt appreciated by your employees). The bottom line is that I will not be able to meet my safety goal, no matter what, by using these measures.

This is the classic case of misuse of a macrostatistic at a microlevel. Even with a single injury, any supervisor in this scenario cannot meet their goal.

As management, we have just destroyed any supervisor's incentive to focus on safety as a job duty. They will have to look elsewhere for that bonus. Perhaps in pushing production numbers to the extent that corners would be cut that impact safety in the workplace negatively.

Sadly, despite the fact that these macro numbers do not translate well into use at the department level, they are still being used. Why? We have grown used to the "traditional" metrics and seem reluctant to embrace new sensible ones. Leading indicators offer a way out of this trap.

Instead of focusing on the end product, the loss, focus should be given to what actions can be taken to contribute to the risk controls that can impact loss. This should come from the safety business plan.

SAFETY BUSINESS PLANS

In this chapter, I have made remarks about the need for alignment with an organization's safety business plan and the need for leading indicators to be aligned with that plan. The development and structure of safety business plans are thoroughly covered in my book *Managing Risk: Not Safety*. Therefore, I will focus on how leading indicators work hand-in-hand with a safety plan.

Leading indicators can be developed without a safety business plan, but to do so, makes the effort haphazard. The plan should cover safety actions, expenditures, and projects that are anticipated as part of the risk control program. These projects might also include capital expenditure plans that need to be implemented over the course of a year. Or, in other words, the very things we say are good leading indicators.

It should be noted, however, that every component of a safety business plan does not have to be covered by a leading indicator. As previously discussed, it is good business to have to have those items identified as impacting KPIs as part of the safety business plan and measuring them. However, there are other components to a safety business plan that are there to identify and implement controls that may not be addressed with leading indicators.

Taking the example of the supervisor in the previous scenario, let us explore how we could expand our metrics in measuring their performance. Such measures can be adopted as instead of punishing supervisors for incidents occurring in their departments, reward him/her for carrying out components of the safety plan. In this example, the supervisor's bonus structure would look more like the example in Table 5.1.

What we now have are safety actions that are in the control of the supervisor.

Unquestionably, this is one of the best uses of leading indicators in the world of metrics. However, even in this use, we cannot simply say that we did it or did not do it. We still need a measurement for the supervisor that identifies their completion, or failure to complete a task. The measurement should also be flexible enough that it gives partial credit so that you do not have a pass-fail system that is too draconian in nature.

The way this is normally accomplished is to track a percentage of completion of the assigned tasks. This then begs the question of if we need a "check and balance" system that measures the lagging side of the formula. The answer to this question is an unequivocal "Yes." As in our earlier fleet example, it does me no good to have my leading metrics achieving 100% success if my plant incident rate is either not improving or, worse, going up.

Consequently, it is not uncommon to see the earlier example expanded as shown in Table 5.2.

TABLE 5.1

Supervisor Leading Indicator Safety Plan

Supervisor John Smith Safety Plan

Item	Objective	Due Date	Status
Hazard Communication	Review and update chemical inventory	Q1	Complete
Hazard Communication	Have all new employees trained	Q3	In progress
Inspection/Test	Have all limit switches tested	Q1	Incomplete
Inspection/Test	Have noise monitoring done	Q2	Complete
Procedure	Update LOTO procedure	Q3	Complete
Procedure	Develop confined space entry procedure	Q4	Complete

In Table 5.2 version of the leading indicator measures, the plant manager has decided to include the TRIR and LTA in order to remind the supervisor that even if they meet all their leading goals, this measure cannot get too far out of control. However, because the supervisor is no longer "living and dying" by the lagging rate, they still can get a significant portion of their safety bonus based on the achievement of leading indicators.

This type of mixed grading can also tell a quick tale of if the supervisor is "pencil whipping" goals. If the quality of the leading indicator is not being met, the inevitable outcome is that the lagging rate will not improve. Consequently, if a supervisor repeatedly meets the leading indicator goals but fails to meet the "loss rate" goals, one of two things is wrong. The supervisor is not delivering the leading indicator in a quality manner or the leading indicator selected is not addressing the problem. In either event, the need for a fix is evident.

This once again speaks to the need of lagging indicators combined with leading indicators to create a balanced scorecard that measures using more than one metric.

As a final note on building your list of leading indicators you choose to track, keep this simple axiom in mind. Someone must gather the data, collate it, and report it. That "someone" is usually the safety professional. When structuring your programs, keep this data gathering and report generation in mind. It will provide a sanity check for how many leading indicators you will apply in your final list.

EMPLOYEE SURVEYS

Dan Petersen was a leader in our field of safety. A visionary who saw what the future of our field should look like. He was also a mentor and friend. In his book, *Measurement of Safety Performance*, Petersen promoted the use of the employee survey as a leading indicator.

While I agreed with Petersen on many things, I did not have the confidence in employee surveys that he embraced as a strategic metric. However, in defense of Petersen's view of the value of surveys, he normally dealt with upper management

---

TABLE 5.2
Supervisor Leading and Lagging Indicator Safety Plan

Supervisor John Smith Safety Plan

Item	Objective	Due Date	Status
Hazard Communication	Review and update chemical inventory	Q1	Complete
Hazard Communication	Have all new employees trained	Q3	In progress
Inspection/Test	Have all limit switches tested	Q1	Incomplete
Inspection/Test	Have noise monitoring done	Q2	Complete
Procedure	Update LOTO procedure	Q3	Complete
Procedure	Develop confined space entry procedure	Q4	Complete
TRIR Rate	Meet plant goal of 1.5	Q4	NA
LTA Rate	Meet plant goal of .5	Q4	NA

in his interaction with the companies with which he consulted. He quite correctly viewed these surveys as a communication tool for employees to anonymously tell this upper management how they viewed the safety culture of their organization.

Viewed in that context, one cannot put up much of a counterpoint to their value. Further to this point, let us concede that surveys hold some value in getting the safety temperature of an organization. Adding to this list, they also provide employees a method to communicate their concerns. Any doubts I express come in two areas.

First, is the current age we live in from a communications perspective. In our current age of information overload, surveys have become ubiquitous in our society to the point where I have come to question their value unless properly prepared, followed up, and evaluated. We will touch on this preparation, follow-up, and evaluation in a moment.

Second, if a safety professional at any level needs a survey to tell him or her what the employees of their organization think of the safety culture, that indicates to me that said safety professional is spending too much time in their office and not enough in the field. Even as I rose to the highest management levels in the company, I made certain not to lose touch with the field. When I say "the field," I mean getting out of the office to the plants, talking to the employees, and creating a grapevine for constant communication.

Once upon returning from a trip to several of our manufacturing locations, the President in charge of that division commented to me that he was impressed with my extensive travel to the plants. What he was saying was "you can't keep in touch with what is going on out there from your office." As safety professionals, this is a golden rule we can never forget, no matter what level of management you attain or what fancy title you carry. If we need an employee survey to tell us what our people are thinking, something is terribly wrong.

My prejudices aside, let us explore some of the positives and negatives of using surveys.

I have found that one of the keys to a successful survey is to keep it concise. Surveys that are too long can result in a "check the box" approach as it progresses since we all have limited attention spans even for surveys. On the other hand, surveys that are too short often don't tell the story. Consequently, planning needs to go into the questions and focus on what are you trying to find out.

Surveys should also be used sparingly. Human nature is such that when we are inundated with something, like we are with surveys, we tend to respond only when we have a complaint that we would like to vent. Otherwise, we ignore them even when our message is that all is good with the world. As I stated earlier in Chapter 4, addressing data, my concerns are around the quality of the data we get from surveys.

SELECTING QUESTIONS FOR A SURVEY

When selecting the proper questions for a survey, we must also allow for a variance in the answer to the question.

Below are the questions of a five-question survey I recently encountered. It was to be handed out to employees for them to fill out anonymously. The first problem is that all of the questions are in a *yes* or *no* format. The first two should suffice to make my point.

1. Are safety policies effective at providing a safe workplace?
2. Does my workplace's safety program address the risks I encounter daily?
3. Do you have all the PPE you need to do the job?
4. Does my supervisor believe in following company safety procedures?
5. Does my supervisor listen to my safety concerns?

While the intent of this survey was a good one, the validity of the results would be questionable at best. The first trap this survey has fallen into is the failure to carefully consider the words in the question structure. For instance, in Question #1, it is asking if safety policies the organization uses are effective.

The use of the word "effective" in itself requires a bit of definition. Furthermore, the survey taker would hardly be in a position in many instances to make that determination. It first assumes that the person taking the survey is familiar with all of the safety policies of the organization. Then it asks for a simplistic black or white answer without any details to what the respondent meant.

Even if we changed the word "effective" to "followed," it is a shorter leap in logic. Now, all the taker has to decide is if the boss is enforcing company policy or has gone rogue. However, that still assumes the taker of the survey has an intimate knowledge of all company policies. Perhaps that should be the first question.

The second question suffers from the same problem. If the answer is "no," what does this tell me? Is the risk we are missing substantial, or is the taker of the survey referring to a minor risk, which we have considered and accepted within our current approach to managing risk?

Perhaps the greatest challenge of any survey is the Yes/No format. It backs the taker of the survey into a black or white awkward position in a world that is filled with grays. There are times when, if the question is worded properly, a yes/no response works. It certainly makes compilation of the results easier, but more often than not we need to allow for a range of responses.

That said, although I never thought of an employee survey as a traditional leading indicator, they can be a barometer of measuring the "perceived" state of a safety culture. I say "perceived" since, as stated in the previous example, it is difficult to validate employee surveys as much more than opinion. However, there is an old adage that states "perception is 90% of reality." In this manner, to some extent, one can get a feeling for where the safety culture stands from employee responses.

Table 5.3 is an example of typical survey questions that can be put to the employee workforce and how you can give some "wiggle room" on responding to the questions.

Of note, in the survey in Table 5.3, there are a mix of yes/no responses that also offer more options on the response. Some of the responses offer multiple choices. The question that is structured begging for clarity is Question #9. This is one of the great sins in a survey question. The person putting this survey together combined two issues in one question. As the person taking the survey, I may have

TABLE 5.3

1. How would you rate the Safety Culture of your organization?

   Very Strong ⬜   Strong ⬜   Average ⬜   Weak ⬜

2. Are the safety procedures you must follow clear or do you need additional guidance?

   They are clear ⬜   I need additional   I need additional clearance
                         guidance ⬜   on a few ⬜

3. Do you see a need for additional safety procedures to make it clear what is expected of you?

   Yes ⬜   No ⬜   It is not a procedure
                                                    problem ⬜

4. Do you feel you have an available line of communication to voice your safety concerns with management?

   Yes ⬜   No Sometimes ⬜   Sometimes ⬜

5. How often is safety brought up during meetings and job briefings?

   Daily ⬜   Weekly ⬜   Monthly ⬜   Rarely Ever ⬜

6. Do you feel your fellow employees care about safety?

   Yes ⬜   No ⬜   Some do, Some don't ⬜

7. Do you feel employees who ignore safety rules are disciplined by their supervisors?

   Yes ⬜   No ⬜   Most are but some aren't
                                                    because of the
                                                    supervisor ⬜

8. Do you feel that your performance review takes safety into consideration?

   Yes ⬜   No ⬜   I am not sure, this was
                                                    not covered in my
                                                    review ⬜

9. Do you feel that you have all the tools and Personal Protective Equipment you need to do your job safely?

   Yes ⬜   No ⬜   Mostly, but we could use
                                                    some improvement ⬜

10. List the training you would most like to receive.

   • _____
   • _____
   • _____
   • _____

(Continued)

TABLE 5.3 (Continued)

11. What do you consider to be the most hazardous parts of your job?

a _____

b _____

c _____

d _____

12. Comments: To improve safety in the workplace what additional suggestions do you have?

_____

_____

_____

_____

_____

concerns with my tools and be happy with my PPE. When we get the response, which is the concern expressed? These two issues should have been asked in two separate survey questions.

The Table 5.3 survey also asked for actual written responses. Most people conducting surveys hate written responses if, for no other reason, the employee penmanship may be so abysmal that you cannot read it. Nevertheless, this allows for more flexibility on the part of the person taking the survey. In the end, you are providing the employee with a better the survey.

However, this brings up the second complaint for written responses. Such responses make summarizing results more difficult. All of which takes us back to our basic question. Are you attempting to create a fancy chart or actually learn what the employees are thinking so that you can apply it to the improvement of your safety program?

Even the handwritten nonstandard responses can be "categorized" through the answers. Table 5.4 is an example of how that was done with the last question for a department survey.

This survey was conducted on a city's utilities department. As can be seen, even though the responses were "written in," a commonality emerged with exposure to traffic. Four of the top five issues were traffic related.

Another consideration of surveys is that we rarely think of middle and frontline management as a survey target. As an example, the survey in Table 5.3 was given not only to the employees in the field but also to their direct supervisors. The purpose was to gauge their sense of where the organizational safety culture stood as well. Comparison of the results of the two could be illuminating to say the least.

SURVEYS AND MEASUREMENT

Perhaps one of the greatest challenges in surveys is how we measure the outcomes. For the most part, the concept of measurement in surveys is to measure improvement or a declining work safety environment or culture. How do we do that?

TABLE 5.4

What Do You Consider the Most Hazardous Part of Your Job?

Rank	# of Responses
1. Drivers not coming to a stop	8
2. Drivers using cell phones	7
3. Confined space entry	7
4. Drivers speeding	7
5. Working in high traffic areas/on streets	7
6. Overgrown easements	5
7. Animals (need for better repellant spray)	7
8. Chemical hazards	4
9. Working on canal banks	3
10. Working in elevated areas/jobs	3
11. Heavy lifting	3
12. Heat stress	2
13. Rowdy patrons	2
14. Driving	2
15. Electrical panels	2
16. Other	8

To validate the survey over time, the same questions would have to be asked repeatedly of the same employee group. To conduct the same survey at different locations would give you a "temperature" for that location but not be valid on measuring the improvement of the initial location.

When developing surveys, as stated earlier, there is a temptation to keep a simplicity in results by sticking to Yes/No questions. The problem, of course, with this concept is that it does not allow for addressing gray areas or forces the survey taker to reply with either an absolute one side or the other. Therefore, surveys that use graded responses, such as a numerical value of 1 to 5 or "Strongly Agree," "Agree," "Disagree," or "Strongly Disagree" type of approach, provide more flexibility. These "scores" can then be taken into account by averaging is there an improvement or not.

As for survey questions, there is no set pattern, but the surveys are naturally attuned to if management is doing a good job. Since the target audience is normally the "boots on the ground" employee, the average question is aimed at getting their view of the safety culture. Typical questions revolve around management credibility and "walking the talk." Consequently, many surveys will have questions such as:

- Does management follow through with safety procedures?
- Does management comply with its own safety rules?
- Does management listen to employee suggestions for safety improvements?
- Does management discuss employee safety in meetings?

Surveys can also drill down into the same questions, but instead of addressing upper management concerns, it can break it down into upper management and then frontline management. This latter approach of breaking down management levels can be important since it can show a disconnect between what the upper management preaches and what the frontline management practices.

This is one area in which a survey can be revealing to upper management who may believe that their "marching orders" are being carried out; when, in truth, they are not.

Surveys can also be more specific by asking drill-down questions in areas such as:

- Adequacy of training
- Adequacy of equipment
- Communication with management
- Practicality of work schedule expectations
- Follow-up on incident investigations
- Safety Committee effectiveness
- Motivational programs

Once again, there are no "set" questions that are asked. Survey questions should be customized to fit the organization.

As a final note, I will reiterate that the first question when developing any survey should be "what are we trying to accomplish." If you want to know if the organizational policies are being carried out in the field, ensure survey questions ask this. If you want to know how employees feel generally about the safety culture, you had best be prepared to define that culture.

LEADING INDICATORS FOR UPPER MANAGEMENT REPORTING

Perhaps the most difficult use of leading indicators is when applied to upper management reports. We will touch on this topic again in later chapters, but as I have already pointed out in an earlier chapter, upper management typically looks for succinct metrics that tell at a glance where they stand. The ultimate example of this is the profit–loss statement. It delivers the message quickly and at a glance.

Leading indicator reports taken to the strategic level do not translate well. In fact, most of my experience of attempting to take leading indicator reports to upper management have been abject failures.

Perhaps we should stop for a moment and define the term "upper management" more clearly.

I suppose when you are the plant safety engineer, upper management, means the plant manager. For someone working at the division, IBU or corporate level upper management will take on a different meaning. This will be the president, CEO, or similar type title.

This latter group are the ones most prone to wanting to see concise metrics that tell the tale quickly. If they desire more intimate details, they will ask for them. But, I believe, we have already amply covered the time constraints under which most of these executives operate.

PRACTICALITY AND SENSIBILITY TEST: PART I

I would like to wrap up concluding with a point I made at the beginning of the chapter. As discussed earlier, leading indicators have faced some serious challenges in becoming an accepted metric within our profession. There are a number of reasons for this, most of which I have attempted to address in this chapter. I will reiterate that perhaps the most commonplace reason for the failure of acceptance of leading indicators has been the lack of thought given behind selection of these indicators. Logic has, in too many instances, been replaced with what seems like a good choice that leads us down the path of the shotgun approach of selection.

Previously, I provided a seven-item list of what we should look for in our leading indicators, which I will present again:

1. Does it impact the success of my safety business plan?
2. Has management endorsed this approach?
3. Can I put a measure to it?
4. Is the measure a number that can be normalized?
5. Is the leading indicator working as desired?
6. Does it play a part in engaging the management team in meaningful activities of the safety business plan?
7. Is management accepting the metric as valid?

What I neglected to put as a precursor to this list was my "Practicality and Sensibility Tests." This scale is relatively simple. On the Practicality Test Scale, we ask ourselves, how easy is it to collect the metric being requested. On the Sensibility Test Scale, we ask ourselves, once we have collected it, what good is it? What is it telling us?

I bring these two scales up for a good reason. In the Preface of this book, I pointed out that much of what you will read in these pages is based on experience, not academic theory. That experience comes from having started my career at the plant level, worked up to a division level, then to a corporate level, and eventually to an independent business unit level. During this entire process, I tried to keep myself grounded and remember my experiences in terms of what worked and what didn't work. As an example, I learned that writing a policy or procedure is not a problem. Being the guy who has to make it work on the front line is a different story at times.

The army has a solution for this. They put you on a staff job to develop SOP and then they transfer you to the field to have to live with the SOP you wrote. A fitting system for injecting practicality and sensibility into any SOP.

Theory often looks like a work of genius until someone has to make it reality. Consequently, I try to use my experience to evaluate what "should work" (and is, in many instances, an excellent concept) against what I have seen work in the field. There is an old expression that goes, "Is the juice worth the squeeze?" We must always ask this of ourselves.

Is my experience sacrosanct? Is it an absolute certainty? Of course not. Experience is the product of our past and to some degree will bias us in certain directions. However, despite those biases, it can be a great teacher. If I reach into a pot of boiling water and burn my hand, experience has taught me that, in the future, reaching into a pot of boiling water will have negative consequences. We must learn to temper our experience with flexibly. But flexibility built on sound reason.

My Practicality and Sensibility Scales are barometers for this approach. We must never stray far from putting ourselves in the shoes of the end user. Experience should not be confused with arrogance or being inflexibly struck in the past. While some of my criticisms, which follow, may seem harsh, they are intended to make us think about what we are really asking for in the metric.

We must remember that when a committee, or someone, comes up with a requirement, suggestion, or cutting edge resolution, there is another "someone" who has to make it happen. In the case of safety metrics, that "someone" is usually the safety person on the front line. If I ask for a particular metric to be collected for safety, nine times out of ten the "collector" will be the safety person. Why do I bring this up?

Earlier in Chapter 3, I discussed the emergence of safety management systems standards. These standards are written with the assumption that if your organization does what they call for in the "systems approach process," your payoff is a safer workplace. However, how do we measure a safer workplace?

Putting additional pressure on the end user of the systems approach, these standards call for self-evaluation on the part of the user to verify things are improving. How do you measure improvement?

Finally, as I will cover in a later chapter, metrics are of inordinate value when I want to make a "sale" to upper management. If I want to sell the systems approach to upper management, I had best be prepared for the most obvious question. "What do I get for this?" I had better have something better than an improved TRIR as the answer.

Add these up and we have a serious measurement problem emerging for systems standards. Ironically, the management systems standards are inspired by, and to some degree built on, the Deming quality approach – an approach that made great use of metrics to measure quality improvement. We will have to learn to do the same. In order to do this, we will undoubtedly make great use of leading indicators as a measurement tool. We will have to choose these measures carefully or we will find ourselves going down the "shotgun" path of failure.

In an effort to demonstrate this looming problem, I will list some select elements of one particular management systems standard's implementation guide. This guide, while well-intentioned, can demonstrate the challenges we will have before us as a profession in selecting good leading indicators as measurements.

As a note, almost all of the suggested metrics in this guide are leading indicators. The sections selected (which are not all-inclusive, but limited for the sake of brevity) are listed in the bullet points as follows:

- Management Leadership
- Worker Participation
- Management of Change (MOC)
- Contractor Safety
- Incident Investigation
- Audits

This is not the entire list of suggested metrics, but it will suffice to demonstrate my point regarding how carefully we must consider our selection of metrics.

The following suggested metrics are taken directly from the standard's advisory section on compiling metrics.

MANAGEMENT LEADERSHIP

Under this section there were 16 examples, all leading indicators. Some of them were as follows:

1. Number of management walkarounds a month
2. Conservative decision-making principles are applied in making decisions about the operations safety of the plant
3. Promoting organization-wide safety awareness via activities such as newsletters, alerts, and toolbox talks

Applying my Practicality and Sensibility Test, let us look at these.

Let us start with #1. This leading indicator is usually offered up in this form or amended to add the amount of walkarounds that a plant manager makes in which safety is mentioned. In either case, they grade low on the Practicality and Sensibility Tests. How exactly do we gather the information on how many management walkarounds take place in a month? Does the Plant Manager and all of his people submit evidence of such walkarounds to the safety department? Do we hire Stalkers R Us? How do we gather this information? Furthermore, even if we did gather it, what do I do with it? Do I report back to the Plant Manager that he made 15 walkarounds last month as opposed to 25 the month before? Does this mean he is falling down on the job? What is my normative approach? Walks per month? Worse if we throw in discussing safety as a requirement how do I track this? As we discussed earlier in this chapter, leading indicators should be carefully focused on specific objectives. Is this focused? Or the shotgun approach?

We must also separate a good metric from a good management practice. Is it a good management practice for a Plant Manager to walk around their plant every day? Of course. I don't know that I ever met one who didn't do so. Those who would stay penned up in their offices certainly would not keep their jobs very long. Nevertheless, while this is an excellent management practice, it is a nightmare to measure. Both the gathering of the data and the end product would present considerable problems in compiling.

WORKER PARTICIPATION

This section had seven recommended metrics. All of these were also leading indicators.

1. Number of safety improvements submitted by workers
2. Number of prejob assessments completed
3. The extent to which personnel consider safety as a value that guides their everyday work
4. Percentage of employees that sense control over their work

Applying my Practicality and Sensibility Test, let us look at these.

Collecting the number of safety improvements submitted by workers is not a problem. It would pass the Practicality Test. However, what does that tell me? What if they submit none. Does that mean apathy on the part of the employee workforce, or does it mean that we are so good in addressing safety issues that there is no need for improvement submissions. Worse, what if I begin to grade my supervisors on how many safety improvements are submitted by their people. Consider the self-fulfilling prophecy that will occur next. The supervisor is not going to be at the bottom of the list for a lack of submitted improvements by his people. So where does this put us on the Sensibility Test? Ditto for the prejob assessments. How long before this is a "forced" measure?

Recently, I went through a series of plants for one company that had a daily prework inspection checklist. The checklist was two pages long and was identical for every department. I would find boxes checked in departments that did not have the safety hazards identified on the list. What this told me was that the forms were being "pencil whipped" instead of fulfilling the intent of the leading indicator. A good concept, but not sensible for the objective.

Moving onto #3, exactly how do I gather "the extent" of a value in someone's mind? Finally, on #4, how do I collect the percentage of employees who "feel" they have control over their work? Certainly I can use surveys for the last two, but would I have hard data or perception surveys? What one employee views as having control may not be viewed by another in the same context. Further to this point, what if the organization's management approach is that they just want employees to follow SOP. They don't want them "taking control" and doing what they want. While this old "command and control" management style might not be in vogue, for some organizations they may be happy with the effectivness of their chosen managment system. Where does this leave me on the Sensibility Test? With more questions than answers?

MOC

1. Approved MOCs
2. Number of MOC change requests
3. Risk assessment is done for organizational changes
4. The effects of the implementation period to organization practices are monitored during the change
5. Workers participate in MOC reviews

Applying my Practicality and Sensibility Test, let us look at these.

Number 1 is easy to track, so it passes my Practicality Test, but leaves us with no normative reference, thereby failing my Sensibility Test. What is this number seeking? The quantity of approved MOCs? If this is the objective how many is enough? If I have ten this month and none next month, what does that mean? Does that mean there were no reviewed changes or we just decided to change without doing them? Ditto for #2.

Number 3 suggests we have now moved into the HR realm and will do risk assessments of organizational changes. While I am sure there are some rationalizations for this, I can imagine the response from your average Plant Manager when the safety department informed him or her that there would be no management change until a risk assessment has been conducted. This fails both tests.

Number 4 fails both tests. I would be hard pressed to collect this info, let alone know what to do with it once we had it.

As for #5, I happen to think this is a great idea. One which I have seen work successfully in many design reviews. However, this is hardly a metric as much as it is a design review procedural issue.

CONTRACTOR SAFETY

1. There is a process for purchasing outside work
2. Types of near misses reported
3. Safety culture of contractors

Applying my Practicality and Sensibility Test, let us look at these.

Number 1 is not so much a metric as it is a procedure. As for #2, it would require a great deal of diligence on the part of the contractor to report these, especially the bad ones, so it fails the Practicality Test. As for #3, this is the classic example of casting a net so wide that the metric is only attainable in a survey. Perhaps the best approach to measuring a safety culture is in determining a contractor's compliance with a preset safety program. An example of this is given in a later chapter of creating a balanced scorecard.

INCIDENT INVESTIGATION

1. A system for reporting and analyzing incidents is implemented
2. Competent personnel conducting investigations

Applying my Practicality and Sensibility Test, let us look at these.

Number 1 assesses whether a procedure is in place or is not in place. Once again we are dealing with sound management techniques that are not lending themselves well to leading indicators unless they are broken down into subcomponents to measure. A procedure being implemented can get checked in the "yes" column whether it is a good procedure or a bad one. To apply a qualitative approach to this question, we must measure what constitutes a satisfactory incident investigation program.

Number 2 is also essentially a qualitative question that would need definition of what constitutes competent personnel. However, once I define it and place a number on it, what would I do with it? It would fail the Sensibility Test.

We could go on but I believe I have made my point. Senseless "theoretical" metrics just add to the difficulty of having leading indicators embraced by upper management as a form of metrics.

The issue with the challenges of measuring management systems will come up again in Chapter 10 when we discuss measuring such systems on a global scale.

In summary, this is not a critique of the management systems approach. It is a critique of simply coming up with a list of things that, in some instances, cannot be easily measured and secondarily will, in many instances, provide us with data that more blurs our message than clarifies it.

It is incumbent on us, the safety professionals, to apply the Practicality Test and Sensibility Test to the selection of all metrics, especially leading indicators. Leading indicators can be a valuable tool or an exercise in futility. Make them the former, not the latter.

REFERENCES

The Good, The Bad, The Ugly. United Artists. 1966.
ANSI Z10 Metrics and Measurements Guide. American National Standards Institute.

6 Results-Based Measures; aka Lagging Indicators

The lagging indicator is far from dead, it is just getting started.

C. Gary Lopez

I am not certain who gets credit (or blame) for coining the term "lagging indicator." I say "blame" because, unfortunately, the lagging indicator (often referred to as a trailing indicator) has, as of late, gotten a bad reputation in our field.

This is unfortunate, because the lagging indicator is neither going away nor is it a poor metric. In fact, as pointed out in the previous chapter, without it, we cannot even validate leading indicators. Despite the negativity that is being associated with the lagging indicator, safety professionals are going to find themselves increasingly using these indicators and in different ways from how they were used in the past.

In the next chapter, I will delve into the concept of value-based metrics, or, more simply put, using dollars and cents to measure success or failure. Nearly all of these metrics will be "lagging indicators" or the term I prefer – "results-based" measurements.

Results-based metrics that "lag" are commonplace in any area of a business. As I have previously pointed out, the most common is the profit and loss statement. What business in the world would view this "lagging indicator" as a bad thing? In fact, it may be the definitive metric for every for-profit organization.

Similarly, in the safety field, we must learn that a lagging indicator is not a dirty word or an out-of-date metric. It is simply one more measurement we use and a common one at that.

What we will have to learn to do is use a greater scope of lagging indicators and select the ones that make sense to addressing risk and loss for an organization.

TRADITIONAL LAGGING INDICATORS

Where did the lagging indicator get such as bad reputation? The safety profession has been using them for years. Why all of a sudden are they undesirable? I submit that much of this bad reputation deals with the frustration of our profession using one dimension metrics. It goes back to our inability to expand our metrics beyond our old "Body Count" metrics. We compounded this problem by using abstract numbers that do not directly reflect the cost of such losses. We imprisoned ourselves in the past with metrics that while still of value, are limited in telling the story of what we do.

As pointed out earlier, the traditional lagging indicators that have been used in the safety field for decades have been focused on employee injuries and illnesses.

The most common of these is the ageless lost time accident rate, aka LTA. The LTA rate at one time was the gold standard of measuring safety performance within an organization. The LTA rate was a product of the old ANSI Z16.1 standards that were developed in the 1930s.

While definitions for judging severity of an incident have been tweaked over the years, a safety professional in 2020 who traveled back in time to 1920 would probably find that their counterparts from back then were using the same definition and formula for lost time accidents. An accident is "lost time" if the employee cannot make it to work the following day of the injury. This rate was the main measuring stick of "success" in a safety program up until the 1980s when two changes occurred that heavily impacted the rate.

As a note, for international operations the LTA still carries a great deal of weight since the definition of what constitutes "recordable incidents" varies from country to country.

The first change was the birth of the Occupational Safety and Health Administration, (OSHA). As noted earlier, OSHA departed from the Z16 formula and leaned on a new formula and determination of what was a "recordable" incident based on Bureau of Labor Statistics interpretations. As the OSHA statistics worked their way into the metrics of organizations, the "OSHA recordable incident" began to slowly displace the LTA rate as a number to be watched. Today we refer to these OSHA recordable rates as either the Total Recordable Incident Rate (TRIR) or the Total Case Incident Rate (TCIR).

The second change that hit the workplace at roughly the same time was even more impactful in changing the "value" of the LTA rate.

THE CHANGE IN INCIDENT CASE MANAGEMENT

In the late 1970s and early 1980s, there was a dramatic shift in how workers' compensation cases were managed. The traditional view in the past was that unless an injured employee was at 100% physical capability to engage in their work duties that they would stay home until they were fully recovered. However, several studies endorsed by insurance carriers demonstrated that an employee that continued to come to work, even if they could not perform all of their job duties was more likely to return to full duty in a timely manner then an employee who stayed home and got used to not working. This created the world of "job restrictions" as a method of managing injury cases. Simply put, job restriction was a system that brought employees back to work who were not fully recovered to do their jobs in a limited capacity or another job that was within their restrictions.

As an example, if I had a job description that required that I lift 50-pound loads and the medical restriction was for me to lift no more than 20 pounds, in the past I would have been told to stay home until I had recovered. Under the "job restriction" approach now, I would be brought in to do other work that did not require me to lift over 20 pounds. This approach was further enhanced to "filter out" job tasks from a job description. In this example, I might be an electrician who due to my injury was placed on a restriction that said no climbing of ladders. My employer could choose to bring me in and make sure all of my normal duties resumed except assigning me any jobs that required climbing a ladder.

A less common use of this system was to transfer an employee to a job that was within their restrictions. For example, returning to our 50-pound. restriction example, I might be transferred to the warehouse to take inventory, thereby bypassing this restriction since the heaviest thing I would have to lift is a clipboard or an iPad.

Initially this approach was met with great resistance by operations management who did not want the burden of dealing with an employee who was less than fully capable of doing their job. This was especially the case when some job transfer cases took on the appearance of "make work" jobs that were not really jobs but a dumping ground for injured employees. A situation that was not good for the morale of either management or the employee's fellow workers.

However, the debate was resolved in the same way many of these types of changes are settled. With the rise in workers' compensation insurance costs, especially in the medical arena, more and more organizations began to see the "wisdom" of adopting restricted or light duty policies in their workplaces. Insurance carriers encouraged it and considered it a critical part of any safety management program. In summary it was reflected in the rates that the underwriters developed which difinitively settled the debate of the wisdom of such systems.

The end result is that today I doubt you could find an organization that does not have a "light duty" or "restricted work" policy, whichever name you choose to give it. Restricted work arrangements are so commonplace that even the treating physicians have been trained to provide restrictions on diagnosis of injuries instead of determining what their work status will be. In the past, it was not uncommon to see an employee return to their employer with a medical slip from the doctor with language such as "take a week off" as the diagnosis. That sort of response would now be sent back to the doctor with a request for restrictions, not work status.

Whether one believes in the restricted work policy or not, one thing became evident. The LTA rate started to take a steady drop downward. Unfortunately, much of this drop was not just better management of safety as much as it was better management of the injury after it occurred.

Once OSHA understood what was happening their response was the creation of the Days Away from Work Restricted, Transfer Rate, or as it is known the DART rate. OSHA hoped that the DART rate would more realistically captured the severity of the incident much like the old LTA rate had been designed to identify. Since the DART captured both the old "lost time" and the new "restricted duty" incidents both were once again being captured.

As one stands back and looks at what the various rates are actually capturing, or attempting to capture, it becomes clear that the "severity" of the incident is what is being targeted. If one reviews the chart in Table 6.1, it becomes apparent that the severity pecking order going from least severe to the most severe is:

- Compensable Rate
- Recordable Rate
- DART Rate
- LTA Rate

TABLE 6.1
Multiplant Lagging Indicators

Location	Compensable Rate	Recordable Rate (TRIR)	DART Rate	LTA Rate	NAICS Rate	Corporate Goals*
Plant A	14.3	7.2	3.9	1.1	8.1	5.0
Plant B	10.3	5.6	5.6	2.2	8.1	5.0
Plant C	13.1	3.3	.8	.8	8.1	5.0
Plant D	5.1	4.2	1.7	.5	8.1	5.0
Total	11.3	5.0	3.1	1.0	8.1	5.0

Ironically, at one time there were several methods of capturing a "Severity Rate" measurement. One method that was used in the old ANSI 16.1 standard was "scheduled charges" for particular losses. As an example, if due to an amputation you lost certain fingers or toes there was "scheduled" days of loss. The illustrations in Figure 6.1 demonstrate how this was calculated.

While this approach today is almost laughable, at one time it was the gospel on how we measured severity.

As we became more "contemporary" in our thinking, the new formula for measuring severity was to count the lost days and apply them using the same formula we used for incidents:

Number of days lost x 1,000,000 ÷ man-hours worked = Severity Index

For some time this "severity rate" became very popular in measuring he severity of an incident after the Z16.1 "schedule system" went away.

However, as one can imagine with the rapid demise of the LTA rate and the rise of the restricted duty approach, the measuring of severity days lost became blurred

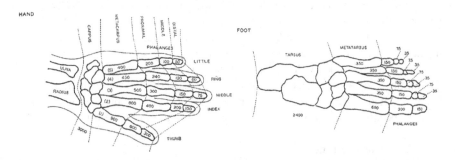

FIGURE 6.1 Old ANSI Z16.1 Hand Loss Chart

to the extent that to identify them became a moot point. Consequently, it was virtually abandoned by most organizations.

This then led to OSHA's increasingly pushing the DART rate as more representative of an organization's true measure of safety performance over the historic LTA rate.

Taking that argument a step further one could say that the TRIR rate is the most representative, except that we end up back to the realization that all TRIR incidents are not the same in severity.

The ultimate solution for this severity dilemma is to introduce dollar rates into the metrics to track the cost of incidents. A topic we will tackle in the next chapter.

USING COMPARATIVE DATA, UNDERSTANDING BLS, NAICS AND SIC DATA

One of the most prominent benefits of the traditional numbers (lagging indicators) is in setting goals and comparative analysis. Before one sets a goal, one must have a benchmark to measure against. Certainly, this can be accomplished using internal numbers, but in many instances the question is: "How are our competitors doing in this metric?" When it comes to comparing how one is doing against similar organizations the TRIR and LTA are the easiest numbers to find for other similar type organizations. This is done by going to the North American Industry Classification Codes, otherwise known as the NAICS Codes.

For the benefit of those born before 1980 the codes used before 1997 were the Standard Industrial Classification or "SIC" Codes. The SIC Code system was introduced in the 1930s to classify organizations on the type of activity or manufacturing process they were engaged in as a business. In this manner, groupings of businesses such as construction, manufacturing, hospitality businesses could be classified. As a note, these classifications were not simply for safety and health recordkeeping purposes, but primarily for other reasons.

In 1997, it was felt that the basic three number SIC system did not accurately identify the subsectors of a grouping (such as manufacturing). Consequently, the more elaborate NAICS code system was introduced to expand on the old SIC system. For example, if you are dealing with manufacturer of say cement and concrete products the code will start out with a basic four digit code and tunnel down to a 6-digit code for specific sectors of that industry. Therefore, while the whole category of cement and concrete manufacturing is NAICS 3273, if your company only manufactures concrete block and brick you can tunnel down to NAICS 327331. While seemingly a nuance, what this does is gives an organization the ability to identify more accurately comparative rates for their specific line of work. As evidenced in Table 6.2, in this particular example, the TRIR for all concrete and cement manufacturing is 4.4. For the subcategory of block and brick manufacturing it is 7.5 (based on 2018 BLS numbers).

While we can debate the merits and shortcomings of each system forever, the point is that the NAICS System is the one used by the Bureau of Labor Statistics to

TABLE 6.2
Safety Plan Alternate

Incidence rates[a]of nonfatal occupational injuries and illnesses by industry and case types, 2018

Industry[1]	NAICS code[b]	Total recordable cases	Cases with days away from work, job restriction, or transfer				Other recordable cases
			Total	Cases with days away from work [b]	Cases with days of job transfer or restriction		
Cement and concrete product manufacturing	3273	4.4	2.9	1.6	1.3		1.5
Cement manufacturing	32731	2.6	1.4	0.6	0.8		1.2
Ready-mix concrete manufacturing	32732	3.9	2.6	1.7	0.9		1.3
Concrete pipe, brick, and block manufacturing	32733	6.3	4.2	1.8	2.4		2.1
Concrete block and brick manufacturing	327331	7.5	5.3	2.3	3.1		2.1
Other concrete product manufacturing	32739	4.7	3.2	1.6	1.6		1.5

identify safety segments. By accessing the BLS Injuries, Illnesses, and Fatalities tables one can see the Total Recordable Rate, LTA Rate, and Restricted Duty Rate.

These BLS tables are the most easily accessible industry comparative rate statistics available to the general public. They are frequently used for setting organizational goals, since it is the most easily obtained comparison they can get to similar organizations in their NAIC.

Putting the aside the already discussed shortcomings about using these rates as a narrow view of an organizations safety efforts many organizations find this data useful. There is no denying that this listing provides an excellent statistical base to compare your organization to others across the country.

Before the reader falls in love with these numbers, I will repeat my earlier "buyer beware" caveat that goes with these figures. As I stated in an earlier chapter, where data comes from and how reliable that data is must always be questioned. With the new OSHA electronic recordkeeping requirements, this data should improve dramatically. However, in the past the data used for these statistics was based on an annual "sampling" of industries, not a complete data dump of every organization in a specific NAICS category. However, the BLS data has become gospel since it is the easiest to access, and quite frankly for the numbers collected is probably the best we have to reference.

USING NAICS RATES TO SET GOALS

In a previous chapter we covered the "normalizing of data" so that it would be more useful to management by "leveling the playing field" for various operations. There are many different routes that can be taken to normalize the data, however, the most common approach is to take the formula used for calculating LTA and TRIR and slightly amend it to calculate other types of incidents.

In Table 6.3, we have an organization that has decided to capture several "rates" in its reporting system. The rates they have settled on are:

- Compensable Rate
- Recordable Rate

TABLE 6.3
Multi-Plant Lagging Indicators

Location	Compensable Rate	Recordable Rate (TRIR)	DART Rate	LTA Rate	NAICS Rate	Corporate Goals*
Plant A	14.3	7.2	3.9	1.1	8.1	5.0
Plant B	10.3	5.6	5.6	2.2	8.1	5.0
Plant C	13.1	3.3	.8	.8	8.1	5.0
Plant D	5.1	4.2	1.7	.5	8.1	5.0
Total	11.3	5.0	3.1	1.0	8.1	5.0

- DART Rate
- LTA Rate

The logic behind these rates is that they capture the entire range of employee in-juries/illnesses happening in an organization. However, the one rate that frequently raises eyebrows in safety professionals is the Compensability Rate. So let us examine the logic for this rate.

Anyone who has dealt with workers' compensation and OSHA recordkeeping is aware of the fact that the comparison of the two have some common ground, but are essentially like comparing apples and oranges. Workers' compensation cases will throw a larger net over workplace incidents. Many are purely diagnostic in nature and consequently would not be OSHA recordable. Nevertheless, some organizations like to use this rate since it serves as their method of tracking "near-miss" incidents.

For those who have tried to track near misses it can often be an exercise in futility. It certainly is an exercise in inconsistency to attempt to capture them all. Consequently, some organizations take the approach that a "compensable" incident is "theoretically" a near miss. Something happened to generate the compensability. Even if the outcome was purely diagnostic in nature, it represents an event that took place that could have been an accident.

A second rationale for this rate is that it accomplishes capturing the old first aid log experience. In the days of large manufacturing facilities, a plant nurse was a fixture as part of the staff. The plants had a dispensary where employees went for minor injuries. These injuries despite only being first aid were captured in a First Aid Log. In some instances these cases were reported as a company statistic. One might look at them as the forerunner to the near-miss reporting concept.

Over the years, the plant dispensary and plant nurse have diminished in manufacturing settings. As a result any first aid rendered on-site is not easily tracked as it was in the past. In the modern workplace, first aid is normally either done on the job by a first aid-trained employee or handled by an emergency room or urgent care center. Whether it is an emergency room or similar urgent care treatment center there is normally a fee associated with the treatment. This is captured (hopefully) by the workers' compensation system and adds one more rationale toward tracking these incidents.

Those that are treated internally by employees are normally lost to sands of time.

Table 6.3 demonstrates how the company has settled on these four metrics. It also demonstrates how the NAICS code is used as the "industry target" and how an internal target has been set below the NAICS code.

One can readily see from the above chart that four things have been accomplished using the traditional LTA, TRIR and other rates:

1 The rates of various locations have been published in a normalized format
2 The organization management has identified what it is watching to measure "success" in the safety efforts
3 The corporate goal has been established against which each plant is being measured
4 The NAICS Rates of similar organizations are published demonstrating how the organization is doing against its competitors

As one looks at Table 6.3 we see that although there are several rates there is only one corporate goal. Similarly, there is only one NAICS Rate. However, it is not unusual to run into similar charts in which the plants are in different NAICS categories, thereby changing their rate.

The setting of the TRIR as the "primary goal" is not unusual in that the setting of to many goals makes assessing the success at the end of the day more difficult. For example, assume we miss the TRIR goal by an order of magnitude but beat the LTA goal. Do we give praise, or punishment?

Focusing on one goal is normally the simplest measurement of success. The question then becomes on which goal do we focus?

MULTIPLE SITE REPORTING

As one will note from Table 6.3, it reflects not just one location but several locations. Experience has shown that this type of reporting creates another management tactic that is hardly new but works as well now as when it was first conceived.

By comparing the various locations with normalized data, a "peer pressure" among the various locations has been created. No one in a management position wants to be at the bottom of the heap for any rating of any type.

Although in this particular case the "rate" being selected as the corporate goal is the TRIR, it could just as easily be LTA or DART.

Most organizations set goals. Some of the goals are given a higher value or priority. These higher priority goals are normally classified as Key Performance Indicators (KPIs). In a later chapter, we will explore working with upper management to set these goals and how they should interact with other corporate goals.

Selection of the TRIR will give this number the undivided attention of the plant manager. No plant manager wants to answer why they are at the bottom of the pile.

While this peer pressure is felt less in single entity organizations, it still exists when measured against goals.

This drives us to ask the same questions. Are we measuring the right things to identify success? Are we measuring enough to identify success? But perhaps the most compelling question is if the metrics we are selecting are having a negative effect on the decision making of our management in the workplace where safety is concerned.

Let's explore this conundrum in the following example.

GARY'S PHARMACEUTICAL PLANT

You have a pharmaceutical plant. I have two problems that I have uncovered in my risk assessments of the facility.

Hazard #1: You are suffering cuts to the hands of your employees on your packaging line due to a new packaging method that requires that they occasionally shave flashings off of blister packs. Some of these cuts are deep enough that they require sutures for treatment, thereby making them "recordable" injuries. You have tried everything from protected blades to cut resistant gloves, but nothing is working. You are still suffering cuts (some of which require sutures), and consequently are recordable on your OSHA Log.

Hazard #2: You have a tank farm that supplies flammable raw materials for your production process and a recent study by your corporate engineering department found that the grounding and bonding system may not be adequate. Weekly you offload toluene (which is a static accumulator) into a tank farm. The corporate engineering department has advised that the grounding and bonding system be upgraded and a routine maintenance program put into effect to guarantee the efficacy of the system.

THE MERIT REVIEW SYSTEM

Your annual bonus system has 25% of the bonus amount graded on "safety performance." This safety portion of the supervisor's (and Plant Managers) bonus amount is based on TRIR rate. The sutures are directly impacting this annual bonus due to the fact that the recordability of these incidents is raising the plant TRIR despite the low severity of the incidents.

Exacerbating the situation is that the entire corporation is on a "Zero Accident" campaign. One that corporate safety "sold" to upper management.

The tank farm fire risk is not tied into any metric whatsoever. Making matters worse there has not been a tank farm fire in the entire history of your site. As you investigate you find that there has only been one other tank farm fire in the company. Of concern is that the cause of the fire was an inadequate grounding and bonding system.

In order to bolster your case for tank farm captial improvements you question who would pay for a tank farm fire incident. To your dismay you discover that since this is a property matter and not an employee safety issue that technically this is handled by the insurance department. They actually question why you are involved in this matter.

In summary, the plant manager's safety bonus is not impacted by the tank farm loss, nor are his frontline supervisors who are also being graded for their safety bonuses on the TRIR as well.

What we have here is a real-world example of what can drive decision making for what represents "safety" down certain corridors while leaving potentially higher risks unaddressed.

If the CEO tells the Business Unit president, "I am measuring you on your TRIR," why wouldn't the IBU president say to the plant manager, "I am measuring you on your TRIR," and so on and so forth, down the management pipeline. This is the classic management system. You go down the chain of command and measure your subordinates on what you are being measured on that represents success.

Once again we face our old friend the "Body Count" dilemma. We have "sold" the concept that safety is about injuries and illnesses only. If it wasn't, why wouldn't we have more diversified metrics?

However, if this location were to suffer a terrible fire and explosion, even one in which no employees were injured, would it matter that the safety record of employee injuries and illnesses was excellent? If this fire and explosion crippled the manufacturing process, would that not be a significant risk that should have been addressed?

In the beginning of the chapter, I indicated that lagging metrics were not the problem. The problem was more a matter of the lagging metrics we have selected. When I say "we" I mean the safety profession. We have clung to the same single dimension metrics for far too long. Once again, this is not to suggest we abandon them. However, we must expand them to truly capture the role we play in protecting our respective organizations from substantial risks.

In the last chapter I referenced the poor selection of leading indicators when using management systems standards. This applies to lagging indicators as well.

PRACTICALITY AND SENSIBILITY TEST PART II

In the last chapter, we discussed the Practicality and Sensibility Test being applied to the selection of leading indicators as metrics for management systems standards. The thoughtful selection of these metrics is not just a leading indicator problem; it can be equally problematic for lagging indicators as well.

As we explore these metrics we find that many are not as much "pure" lagging indicators as they are the suggested measurements for leading indicators. In an earlier chapter I pointed out that all leading indicators will require a number applied to them and a normalization of those numbers for them to work well as measurements of success. This does not change the classification of the leading indicator, it just serves as the measurement of completion. Not the validation of the success of the indicator.

To demonstrate examples of this and how well some of the chosen metrics serve our purpose let us briefly explore some of the following categories.

The categories we will explore are:

• Worker Engagement

- Risk Identification
- Continual Improvement

I will once again apply my Practicality Test, which applies to the difficulty of gathering some of the metrics being suggested as well as my Sensibility Test for testing what these measures are intended to accomplish if we do gather the data.

To spare the reader too much detail I will not explore every single metric, but focus on a few to demonstrate the difficulty of using these indicators in a fashion that lessens their value.

WORKER SUGGESTED LAGGING INDICATORS

1. Absenteeism rate
2. Percent workers receiving recognition for safety and health activities
3. Percent of safety and health related programs with worker involvement

Applying my Practicality and Sensibility Tests, let us look at these.

Number 1 implies that the percent of absenteeism directly correlates to safety. This rate is easy enough for HR to track, but it fails the Sensibility Test. Tying absenteeism into safety performance has been a "theory" for years, but there is no direct correlation that has been found to prove this concept. Number 2 addresses the percent workers receiving recognition for engagement in safety activities. This in turn begs the question of what constitutes "recognition" of the worker. Number 3 is not much better. Worker "involvement" covers a lot of turf. Who tracks this? Both fail the Practicality and Sensibility Tests.

RISK ASSESSMENTS

1. Percent deficiencies on control evaluations
2. Percent of people required to wear PPE reduced
3. Percent of tasks completed per percent compliance with controls

Applying my Practicality and Sensibility Tests, let us look at these.

Number 1 moves to more esoteric measures such as percent of deficiencies in control evaluations. On both the Practicality and Sensibility Test this scores very low. Especially on the Practicality Test where gathering such information would be tedious and to some degree judgmental.

Number 2 addresses the percent of people wearing PPE being reduced. I am not certain how we would collect that data. Perhaps from purchasing, however, moving to the Sensibility Test what would this tell me? I have less employees working? Or perhaps I had jobs that were no longer required to use PPE because of IH test results? Obviously, the conclusion that is being sought is that there are more engineering controls in place. That is a huge leap in logic.

Number 3 suggests percent of tasks completed while complying with controls. This would fail the Practicality Test simply for the difficulty in gathering this data.

CONTINUOUS IMPROVEMENT

1. Regulatory citations per business unit

Applying my Practicality and Sensibility Tests, let us look at these.

I have a hard time connecting the number of regulatory citations per business unit into continuous improvement. Certainly, this is easily trackable, but how this is a systems safety issue puzzles me.

We could continue this demonstration but once again I think I have sufficiently made my point. Metrics that fail the "Practicality and Sensibility Test" are not metrics we want to embrace.

I will reiterate my statement from the last chapter that this is not a indictment of the systems management approach. It is an indictment of our selection of metrics that will resonate with management. We are not gathering these numbers for us. We are gathering them to demonstrate the success and failures of our safety efforts.

That we need to expand our metrics is a given. What metrics we choose to use in this expansion is another matter. Let us now explore one way of expanding our metrics and coming up with new measurements that resonate with our peers within an organization.

REFERENCE

ANSI Z16.1. 1954. *Recording and Measuring Work Injury Experience*. American National Standards Institute.

7 Value-Based Metrics

What You Choose Not to Risk, Reveals What You Value.

C. Gary Lopez

We have already established that, as a profession, we have a mission to address the risks to the things we hold of value. Because our ongoing theme is that we manage what we measure, we must establish a metric that measures the impact of that risk and the cost of any loss we might incur to what we value. So what metric does that best? Why not the metric used for the very cost we incur in a loss. The dollar sign. As a metric, there is no confusion using a dollar as a measurement. I do not need a dictionary, Wikipedia, or an Econ 101 course to look up the value difference between 99 cents and a dollar. I do not need anyone explaining the formula I used to arrive at an abstract number. I do not need any special translation of the metric. It is a metric all of us are familiar with and requires no elaborate explanation.

This proposed use of the dollar as a metric brings us to the premise of this chapter. Why do our primary metrics, the LTA, TRIR, and DART rates take such elaborate explanations? Why, if the dollar is so evident in terms of value, does our profession not use it more as our metric of choice? Why do we not use it as a metric to identify loss, measuring risk, or any other business function we engage in as safety professionals?

There is probably no area of metrics and measurements that the safety profession can use to its advantage more than "value based" metrics. When I refer to the term "value based," what it means in plain English is using dollars and cents as a measurement for safety metrics.

The advantages of using this metric are multiple. The dollar-based metric will provide clarity to those in the workplace with whom we share these metrics. The use of value based metrics can also go a long way toward expanding the role of the safety professional from someone focused on just employee injuries to being focused on total risk. We can use such metrics to validate capital expenditure requests. It is a metric that we have not used to our advantage. It is time to change.

Ironically, it still surprises me that as our profession was conceived largely by the insurance industry, an industry that has no problem using the dollar as a metric. Why we ventured away from this metric and do not embrace it as our defining metric astounds me to no end.

I have already alluded to the need for us to work more closely with the insurance part of our organization and the industry in general. So let us explore this more closely.

THE SYMBIOTIC RELATIONSHIP WITH INSURANCE

DEFINITION OF SYMBIOSIS: A MUTUALLY BENEFICIAL RELATIONSHIP BETWEEN DIFFERENT PEOPLE OR GROUPS

Somewhere in the dim past of the safety profession, there was a slow drift away from the insurance industry. As I stated above, I find this ironic since it was the insurance industry that acted as the catalyst for the birth of the safety profession and movement.

Unfortunately, along with this drift away from the cradle of our origin, the safety profession also drifted from the metrics used by the insurance profession, as well as much of their philosophy.

This itself is a tragedy because of all the symbiotic relationships the safety profession should have with other areas of expertise; none is more pronounced than the bond it should have with the insurance industry, an industry that the safety profession must learn to re-embrace.

The insurance industry has operated under the same business plan for literally centuries, a very profitable plan to be sure. It works something like this:

1. Identify a risk that the customer will realize needs to be insured to protect themselves.
2. Place a dollar value on what it would cost to insure that risk (underwriting).
3. Sell the protection (policy) to the customer.
4. Inform the customer that the premium price can be enhanced in their favor by the customer taking steps to protect their risk.
5. Put as many protective systems in place (as are feasible) so that claims will either not have to be paid or if they will be restricted to smaller amounts of loss due to the aforementioned protective factors.

All right, I realize this is vast oversimplification of a complex industry, but essentially it is not far off the mark and each of us should be able to relate from a personal standpoint. Allow me to elaborate.

I own a home in South Florida. Our "wind damage" coverage (aka hurricane coverage) continues to climb because of the value of our property rising. As a result of this rise in home values, my insurance carrier approached me and presented me with two options.

Option #1: Upgrade the protection of my home with a better roof, new hurricane-rated doors, and better hurricane shutters.

Option #2: Pay more premium.

The capital improvements to my home (because that is what they would be in the business world) represented a significant investment for me. I was looking at $12,000 in upgrades. The question on the table was do I spend the money or do I simply pay the increased premium?

To decide the correct course of action, I did the math. I calculated what I would save on premium versus what the cost of the improvements and calculated the payback. This allowed me to determine if the payback time span was a good one or would stretch beyond my life span. I ran the numbers and from a pure economic standpoint, this capital improvement made sense for me.

Okay, I realize that this does not exactly make me a candidate for a Wharton School of Business award, but you get the picture. I used a metric that clearly laid out my business case.

In the business world, this is called return on investment (ROI). The question we should ask as a safety profession is why are we not applying this same logic, one based on the dollar sign, as the primary metric of our profession? Why instead are we stuck on abstract numbers?

The dollar sign is the international language of business. For us to ignore it in our equations while setting metrics is sheer folly. At times, we might as well be speaking Martian with our various rates we throw out there as our definitive metrics. Unfortunately, there is a second factor at play with our metrics.

In the safety world, we have made the concept of tying dollars into what we do as almost vulgar. We have all heard it. "There is no price on safety." Either we started this rumor or someone else did. In any event, no matter who started it, if experience has taught me one thing in the business world, it is that there is a price on darn near everything.

Our thinking on dollars not mixing well with safety needs to change. We should not view it as vulgar, but simply as approaching and measuring what we do in a more businesslike manner. How could I have made my decision on upgrading my hurricane protection on my home if my sole rationalization was based on a rate of hurricanes hitting the area? Although this might have helped influence my decision, it was not nearly as definitive as seeing the financial implications listed before me. It was the ROI that closed the deal.

To illustrate how this applies, let us return to an earlier example of metrics impacting risk making decisions that we touched on in the previous last chapter. Only this time let us and add dollars to the metrics.

GARY'S PHARMACEUTICAL PLANT: PART II

In the previous chapter, we discussed the problem of Gary's Pharmaceutical Plant. We were dealing with two different risks: one of which, employee injuries, had a metric attached to it and the other of which, the tank farm grounding problem, had no real metrics tied to it. Consider how our business case for a grounding and bonding program would shift if we put numbers to this equation.

As a reminder, the dilemma we were faced with in this scenario was that the employee injury metric was being closely tracked because of how it impacted the safety portion of the annual management bonus.

Let us look at those numbers.

To refresh the reader's memory, the two issues we are dealing with are as follows.

Hazard #1: You are suffering cuts to the hands of your employees on your packaging line due to a new packaging method that requires that they occasionally shave flashings off of blister packs. Some of these cuts are deep enough that they require sutures for treatment, thereby making them "recordable" injuries. You have tried everything from protected blades to cut resistant gloves, but nothing is working. You are still suffering cuts (some of which require sutures), and consequently are recordable on your OSHA Log.

Hazard #2: You have a tank farm that supplies flammable raw materials for your production process and a recent study by your corporate engineering department found that the grounding and bonding system may not be adequate. Weekly you offload toluene (which is a static accumulator) into a tank farm. The corporate engineering department has advised that the grounding and bonding system be upgraded and a routine maintenance program put into effect to guarantee the efficacy of the system. These cost breakdowns are illustrated in Tables 7.1 and 7.2.

Clearly, once we have put the numbers into a value-based approach, the picture changes dramatically. Asking for $1.2 million capital spending to eliminate $3200 in injuries would truly test an organization's will to achieve "zero accidents." The ROI would not be measured in years. It would be measured in centuries.

Conversely, we now see that a hazard that has produced no losses whatsoever has the potential to impact the organization with a nightmarish loss. A loss that if realized could cost multimillions of dollars.

TABLE 7.1
Cost Breakdown for Cuts

Hazard #1: Cuts on Packaging Line

Hazard	Cost of Improvement	Potential Loss
Employee cuts to hands	$1,200,000*	$3200**

Notes:
* Cost of retooling packaging line.
** Annual average losses due to cuts.

TABLE 7.2
Cost Breakdown for Grounding and Bonding

Hazard #2: Grounding and Bonding at Tank Farm

Hazard	Cost of Improvement	Potential Loss
Fire due to poor grounding during transfer of flammables	$200,000	
Annual maintenance of the system	$30,000	
Product in tank farm		$500,000
Tank farm replacement (assuming total loss in fire)		$4,000,000
Potential Business Interruption		$140,000,000
Total		*144,500,000*

As safety professionals tasked with managing the risk of an organization and protecting the organization from just such horrific losses, which hazard would we assign a higher priority after using these metrics?

Additionally, once we attach numbers to the packaging line loss scenario, who would want to make a business case to spend all of that capital for so little return. Zero accident is a noble concept and wonderful goal. However, there are times when the "concept" and the business case collide, times when there is no sound business case for the "concept."

This is not giving up on our mission as safety professionals. We are not turning our backs on employee safety. What we are doing is approaching the managment of total risk from a more holistic view. We must remember that in the tank farm disaster scenario there is potential for employee injuries as well. We are not being cavlier or irresponsible. It is being responsible on how we manage risk. It is acting as businesspeople as well as safety professionals. To accomplish this goal, we need to embrace value-based metrics.

THE GREAT LIES

As I previously stated, I am not sure when and where the concept of "there is no cost on safety" got started. I call it one of the "Great Lies." In the world of any organization, whether it be a for-profit business, government entity, nonprofit or any other organization with a budget, there is still a need for a business mentality that recognizes there is a cost to everything. Unfortunately, as a profession, we have to some extent bought the "there is no cost to safety" premise.

In fact, I submit that we have bought into the argument so deeply that we have shied from attempting to capture many of these costs, and worse taken a stance in many instances that it should not be a factor. It borders on an arrogance to take a position that we are above such mundane values.

Carrying this logic to its extreme, it implies that safety is the single most important element in running a business. Although this may be prominently displayed on signs throughout your site, the veracity of this proclamation is somewhat dubious. If this were true when a company declared bankruptcy, or conducted cutbacks in employees, the last person left to close the doors should be the safety professional. The absolute last person laid off should be the safety person. I believe it is safe to say we all recognize that this is not typically the case.

In my book *Managing Risk; Not Safety*, I point out that our profession is needed the most in an organization's development stage when that organization hits what I call the Critical Mass Stage. The Critical Mass Stage is when the organization has so much to lose that it cannot afford to operate without safety and risk considerations as part of its business plan and logic. What we have to grasp is that it is "part of" this logic, not the entire plan.

Normally, as safety professionals, we walk into the organization when it has matured and has already achieved this Critical Mass juncture. It is a shame we did not witness the earlier stages. It would perhaps give us a new perspective of why we are there and how we earn our paychecks. We are not there solely to interpret standards, or give safety training, or do audits or fill out the OSHA Log. We are there to protect the organization from significant risks. Risks that could heavily impact the organization. To understand what these risks are, we must learn to think in dollars and cents, not incident rates.

Once again, one has only to reflect on what this means from our personal examples. When we started out in life with no assets or anything to lose, the concern for protecting these assets through the purchasing of insurance or other methods was hardly a priority. Why purchase insurance for nonexistent assets? The business world is not different. It is not until they hit the stage when they have much to lose that these matters become part of the business planning.

We need to comprehend that our role is to protect our organization from large loss and put controls in place based on a prioritization of these potential losses. For those who read into this being an insurance matter, it is anything but that. Insurance is one of the protective systems and controls we put in place. In fact, it is sometimes the fallback system for when our primary controls fail. A system that normally kicks in when we have failed to recognize or maintain proper controls to address our risks, or these controls have been overwhelmed.

A second issue I point out in my book *Managing Risk; Not Safety* is that our profession has earned a reputation (deserved or not) as being regulatory specialists who do not approach issues in a businesslike manner such as other members of the business team. The American Society of Safety Engineers (now the American Society of Safety Professionals) conducted a survey some years ago that was very revealing about how our profession was viewed. The survey was designed to poll "operations executives" on how they viewed our profession. The responses were not always kind. Some of the criticism leveled at our profession was that we did not think like business people, our measurements were not easy to understand and we used regulations as our justification for what we wanted.

Although I disagreed with some of the conclusions, there is no doubt that there was some validity in the accusations leveled at our profession. Especially when it comes to how we measure success. We should not be afraid of using the dollar sign in our metrics. We should embrace it.

Abstract rates, whether they be LTA, TRIR, NCCI, or any other rate, are meaningless to our peers on a business team. They sound like black magic numbers. The dollar sign needs no translation.

As an example, consider the following two approaches while justifying capital at a business team meeting:

Approach #1

"If we spend the capital dollars requested for safety improvements we will realize an improvement in our Incident Rate from 1.2 to .9."

Approach #2

"If we spend the capital dollars requested for safety improvements we will realize a 10% improvement on loss in this area, which translates into $500,000 and contributes to meeting one of our KPIs addressing reduction in operating costs."

If you are a non-safety professional business team member sitting around the table, which one makes more sense to you? Which one would you clearly understand? More to the point, which one would you support?

When I have made this example in some presentations, the question I sometimes get is "What if we can't meet that 10% reduction? Won't our credibility be shot?" This is the same as saying that you do not trust the measures you are asking for to actually work. The guarantee in this 10% reduction is directly correlated to the homework you did to establish what you were asking for will indeed work. Otherwise, what is the reason you are asking for the capital and improvement? Because it feels good?

If we put proper metrics in place to measure the success of what you asked for, we will see the improvement or we will not. If we do not, it means that you asked for the wrong thing. If the request is for something purely preventative in nature, such as the tank farm grounding system, then word it that way. Do not claim a reduction in cost for what did not happen. Request it to prevent the catastrophic scenario you identified using the numbers that demonstrate the potential losses versus the costs. Good business people understand that sometimes we invest to prevent such losses because of the potential impact to the business. They are not strangers to the concept. But it helps when our metric is a dollar sign.

In either format, you are using numbers that the business team can understand, not the abstract rates that do no transfer into logic.

USING DOLLARS AS A MEASUREMENT

In the last chapter on lagging indicators, I gave examples of how the use of lagging metrics were still of value. Let us elevate our game and think of taking some of these same metrics and applying a dollar sign as the measurement.

Table 7.3 is a multilocation chart I shared in the previous chapter to demonstrate how normalizing metrics and reporting them could actually create a competitive atmosphere among various locations.

Let us take a similar approach but expand the metrics on which I report and assign a dollar sign as the measurement for the normalization of the report.

This result is illustrated in Table 7.4.

What we see is that the metrics have now expanded to go beyond the traditional "Body Count" of measuring just employee injuries/illnesses. Fleet, property, and general liability are included as "measurements of success."

Even with the measurement of employee injuries, the numbers are translated into dollars as a metric, as opposed to traditional rates.

To be clear, the metrics selected in this table and the normalization approach are for illustrating the concept. The specific metrics and normalization approaches will vary from organization to organization.

For example, if an organization has no fleet, then to dwell on these numbers would make no sense. Further to this point, they may like the idea of measuring in a different rate such as cost per hundred miles driven. The specific rate is not as

TABLE 7.3
Multiplant Incident Rates

Location	Compensable Rate	Recordable Rate (TRIR)	DART Rate	LTA Rate	NAICs Rate	Corporate Goals*
Plant A	14.3	7.2	3.9	1.1	8.1	5.0
Plant B	10.3	5.6	5.6	2.2	8.1	5.0
Plant C	13.1	3.3	.8	.8	8.1	5.0
Plant D	5.1	4.2	1.7	.5	8.1	5.0
Total	11.3	5.0	3.1	1.0	8.1	5.0

TABLE 7.4
Multisite Cost and Lagging Scorecard Revise

Location	WC Total Reserve	WC Cost per Man-Hour	GL Costs In 000,000	Fleet Cost per mile	Fleet Losses	Cost per Vehicle	Prop Loss	Prop Loss per M
Plant A	$6815	2¢	$1.5	$.05	$500K	$2.5	$1.2M	$500K
Plant B	$13,294	3¢	$2.4	$.10	$1.2 M	$8.4	$3.4M	$2B
Plant C	$9836	2¢	$.7	$.08	$1.1 M	$5.5	$500K	$800M
Plant D	$3006	1¢	$.5	$.02	$100K	$.5	$125K	$400M
Total	$32,951	2¢	$1.3	$.06	$725	4.2	$17.3M	

important as making certain the rate makes sense, measures what your organization considers "success" and is put in a dollar form that all can understand.

In this example, the workers' compensation normalization approach uses a measurement of "cost per service hour." The logic behind this metric for this organization was not random. It is a metric that was selected because it tied into other similar type metrics used in other departments of the organization to budget costs.

ALIGNING SAFETY METRICS WITH OTHER CORPORATE METRICS

Like any organization, various departments can track their overhead costs based on "employee load." Employee load is the number that is added to the basic salary of an employee to demonstrate their cost to the organization from a budgetary standpoint. As an example, if I am hiring an employee at $100,000 dollars a year and my employee load for vacation, health benefits, and other employee benefits is a "30% load" cost, I must budget $130,000 as the true cost of the employee on my payroll.

Naturally, this spills over into other budgetary areas that impact the cost of operations. One metric that is used to measure this impact is the cost per man-hour, or service hour to use the new term.

In many organizations, this number is closely tracked and tied into labor costs. Consequently, this number could influence the use of temp employees over new hires. It is yet another metric we should not take lightly and one we can use to address losses for safety reasons. One has to consider only that the majority of non-salaried employees are paid an "hourly" wage. Consequently, if I am adding 10¢ an hour to an employee cost, this may seem like a paltry amount when standing alone, but it can make a difference when one starts adding up employee load. Especially when one factors in that this is a controllable number. It is not some prenegotiated salary or benefit rate that is fixed. This is a number that can be impacted by what we do as a profession.

This is yet another example of how converting from abstract numbers to real dollars and cents makes the measurement much clearer for the other members of the management team to understand and bring safety statistics into perspective.

INTRODUCING DOLLARS INTO INCIDENT INVESTIGATIONS

Perhaps, there is no place that the introduction of dollars and cents into the metrics formula can be more impactful while investigating accidents.

Because our traditional method of measuring loss from accidents has been "Body Count," most safety professionals develop tunnel vision during the incident investigation process. Worse, we then "undersell" the loss because we have convinced our upper management to measure success or failure using these same numbers.

The sad part of this is that in many instances our inability to identify the potential cost of the loss can result in our decisions to not give higher priority to the controls and recommendations we submit to prevent future similar types of incidents.

In an earlier chapter, I described an incident where a forklift driver struck an in-rack sprinkler and released untold gallons of water on to product in our warehouse. Let us examine this incident in greater detail and apply value-based metrics to the investigation.

WAREHOUSE ACCIDENT: PART II

To refresh the readers memory, this incident occurred when a forklift driver was replaced by a driver that was not qualified (licensed) on the particular forklift he was operating. His replacement was not as skilled as he was in operating the forklift nor as familiar with the warehouse layout where he was asked to operate the forklift.

The incident unfolded as follows. During a routine day in the warehouse, the new driver hits an in-rack sprinkler head. The sprinkler head ruptures and water flows freely on finished product and other materials stored in the warehouse. The immediate outcome is that warehouse operations in the shipping area are shut down for a day, whereas the damage is repaired and the mess cleaned up.

From an employee injury standpoint, this turned out to be a minor incident. The forklift driver bumped his head on the mast of the forklift. He was taken to the emergency room to be checked out, but displayed no signs of a concussion and was subsequently released as there were no other injuries.

Cost of the injury was $550 for the emergency room visit (as a note, this was the cost at the time of this accident, I shudder to think what the cost would be now).

As I pointed out in the previous chapter, the initial incident investigation report focused on the fact that the operator was not properly trained. The leading recommendation was to address the lack of training. Our solution would be to enhance the training of forklift operators and reinforce the message that only trained operators are allowed to drive forklifts. While not a bad recommendation, as events would prove this was not even the tip of the iceberg in costs or resolution of the issue.

Using our "Body Count" metrics, this employee's injury was not recordable as a LTA or TRIR and costs only $550. In a sense, one could say, "we dodged a bullet on this one." But did we? Let us examine this same incident using dollars as a metric to track all loss and the resulting impact of the investigation.

As I previously pointed out, this incident was one of the first to be addressed in our "new world" of incident investigations in which we were beginning to track the total cost as a result of an accident. Not just the injury or potential for injury to the employee. What we discovered during the investigation was the following:

Cost of water damaged finished product that is now scrap	$600,000
Cost of cleanup of water damage	$ 45,000
Production downtime in warehouse	$150,000
Cost of damaged product disposal	$ 25,000
Repair of sprinkler system	$ 13,000
Total cost	**$833,000**

I suppose being entirely correct we should say $833,550. We do have the $550 for the employee injury to add into the mix. Not that it would make much of an impression after the rest of the costs were tallied.

As we look at these numbers, we need to ask of ourselves which incident investigation more accurately reflects the real cost? Or perhaps more to the point, which one will get management's undivided attention?

The first argument I normally get when presenting this scenario is that it is unfair to account for production downtime. The employees are getting paid anyhow, why must we charge that off as an accident cost? The reason is because idle employees being paid to stand around is not much different from an injured employee getting paid and not doing their job. Furthermore, we have the nominal

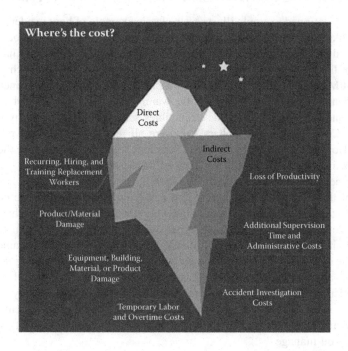

FIGURE 7.1 Accident Iceburg

cost of maintaining warehouse operations whether it is shipping and receiving product or standing idle. The "fixed" costs are there whether you are operating or not.

For years, we in the safety profession have been indoctrinated to the "accident iceberg" (Figure 7.1); however, we do little to pursue the logic of the graphic.

Although we have been exposed to this graphic for years, for most safety professionals, it means very little other than a graphic representation of a general concept. However, the graphic representation should tell us that there is more to an incident than the immediate costs. In fact, as illustrated in the example of the forklift incident just given, these costs are truly representative of the larger impact of the loss. Consequently, it is a "shame on us" if we are not capturing the larger metrics of a loss such as this one.

As we view an incident like this, our roles can change and we have the opportunity to change not just how we do our jobs, but how we are viewed and how metrics can impact how operational costs are measured.

THE SYMBIOTIC RELATIONSHIP WITH INSURANCE: PART II

The aforementioned example illustrates a classic example of how incident costs can become like hidden fees. In this particular scenario, the costs were real but the Plant Manager didn't see them because the corporate insurance department handled the claim for the property damage. As a result of this "corporate" approach to billing, the Plant Manager did not see the numbers, he quite frankly had no concern about the incident beyond the down time of the warehouse. Making the matter worse was that this incident did not meet the deductibles set up with the company insurance policies. Consequently, the dollars lost were all real costs that were paid out of pocket. However, once again the pain of the loss was softened because there was a corporate insurance "fund" for this type of incident. The bottom line was that no one was being held accountable for the $800K.

All of this changed when I created a new system for tracking the cost of all such losses. Losses below the deductible were direct operating costs charged back to the plant. They also factored into an overall charge back system to the individual sites regarding how much their "share" would be when it came time to determine how the cost of insurance premiums were divided up among the various sites. As one plant manager quite correctly summed up the changes: "It looks like communism is dead and that the sinners are finally going to have to pay the piper." He was correct. Suddenly, there was a passionate interest in incident costs by all the plant managers. It was the proof positive of the golden rule, "what you measure you manage."

The impact then rippled to our design system. As the investigation deepened, we were shocked to find that this was hardly our first in-rack sprinkler incident. Worse we had this sprinkler design at a half dozen other warehouses and were about to install them in two more.

This resulted in the question of who authorized the in-rack sprinklers and allowed them to be installed even after several incidents were known to have occurred. What we found was the classic example of how a lack of coordination in management can create loopholes that in turn become problems.

The answer of who approved the in-rack sprinklers to be installed was the "Risk Manager." It was our insurance department that had "approved" these installations. However, the real shock was that they had completely bypassed our strict design screening procedures.

For some industries, design review is a "nice to have" system. In the chemical/ pharmaceutical industry, it is a "must have" system. These procedures are in place for a number of reasons. First, to make sure design changes do not impact the safety of our production processes. Second, to make sure that they make sense from a capital expenditure standpoint by demonstrating some type of ROI, and finally not to impact our quality and validation requirements for our products and processes.

However, in this case for whatever reasons, the fire systems were viewed as removed from this process as they were an "insurance department requirement." This rationale provided a carte blanche free pass from our design and capital review process. Much to the amazement of some of our plant managers, I put an end to that particular "loophole."

This incident was the classic example of why the safety and insurance departments need to be in lockstep. We were simply not talking to each other. This uncovered a problem that many organizations suffer from to this day due to this disconnect between the insurance and safety departments.

As I previously pointed out, in many organizations, the insurance departments often report to the financial department and safety normally reports to operations. This is disconnect step 1. Disconnect step 2 is that due to the approach insurance departments take, there is an opportunity for a different strategy on protection that if discussed with the operations department could result in entirely different approaches.

In this case, operations were not informed of the issue. They assumed that this was a hard and fast requirement as did those running the design review process. The trap was set and sprung.

Once again, this is not an indictment to the people who staffed our insurance department or any other insurance department. They were simply doing what they considered as prudent. However, after this incident, we had 833,000 reasons why in-rack sprinklers were no longer an option in our warehouse designs.

I might add that this is also not an indictment of fire protection measures, in-rack sprinkler systems or insurance company loss control practices. It is about metrics and how metrics can impact thinking in many different areas of loss. Under the right set of circumstances, in-rack sprinkler systems might make perfect sense for some. However, past incidents, potential costs of a future incident, and other solutions that might take away the need for such systems should go into the thinking before pressing the "approved" button.

Until this type of loss was accurately measured in dollars and made visible to our management team, it was all but invisible. Why? Because we did not have value-based metrics that measured the loss and reported it clearly to the management chain of command.

In comparison if it had been a severe employee injury, we would have been all over it with recommendations to make sure it never happened again. Why? Because that is what our metrics have historically been attuned to do. Without a metric tracking the loss, there were no management controls in place.

I am sure there are those who could come into that warehouse and make a compelling case on why the in-rack sprinkler system was necessary. It would have dealt with sprinkler densities, material stored, studies done by excellent fire protection engineers on extinguishing of fires. That would be missing the point. We manage what we measure. In this case, the lack of measurements removed management from its normal decision-making capacity.

Even if it was clearly demonstrated that the density of the sprinkler system was not sufficient to extinguish what was stored in the racks, there should have been a management review of the problem to determine if there were alternatives or a simple acceptance of the added risk.

Before such a system was put in place, two decisions should have been made: first, the additional risk was not acceptable; second, there were no other alternatives but to install the system. Conversely, the management decision might have been that the risk of an in-rack sprinkler head being broken was viewed greater than the density of extinguishing capabilities. In either event, the operative words are "management decision." For such a decision to be made in a sound manner, solid value-based metrics would be necessary.

As a final note, this is not a matter of management trying to evade state of the art fire protection measures. This is management making a business decision based on hard numbers. Numbers such as if there is an ROI, the risks involved in not upgrading the system, alternatives to installing the system and what risks were we accepting. Without value-based metrics, how does one do this? Use tarot cards?

USING VALUE-BASED METRICS FOR ROI

In many instances, the safety professional is placed in a position of asking for capital investments for equipment or facility improvements that intuitively we know makes the workplace safer, but we have trouble putting into the same context as says a production department asking for a new automated process.

For the production department, it can break the numbers down into how many widgets are being made now with "x" amount of work force and how many widgets we could make with the automated process thereby making it a profitable investment. These numbers are crunched by the "bean counters" and they will announce the payback on the investment is good based on the ROI or that it does not make good business sense. Based on this rationale, the capital is approved, or rejected, by executive management.

Now move to the safety professional. We have a more difficult time with the process. The reasons are much the same as with much of our indoctrination and training. Our logic path is not as clear. Our logic path may be "you have to do it because its safety," after all isn't safety "Job One?" Alternatively, we may use other such arguments that ignore any type of financial rationale.

Although this "appeal" may work occasionally, having sat in enough finance meetings, I can assure you it leaves most of our peers on the business team skeptical and questioning our business acumen. Appealing for capital based on either regulation, zero accident, or safety, first philosophies leave fellow business team members cold.

Naturally, there are times when the "regulators say we gotta do it" case is going to be sufficient. As an example, if a Fire Marshall comes in and specifies an upgrade or expansion of the sprinkler system on your new building expansion or you do not get your Certificate of Occupancy (CO), that settles that. You can then fall back on the "regulatory" justification.

However, let us take the same sprinkler system and make it a nonregulatory issue. Let us make it a request from the insurance carrier. In this scenario, your answer needs to be based on logic and a value-based metric approach. One that can demonstrate what is being requested makes sense, or is conversely just an insurance carrier request to enhance our protection.*

*Once again, for clarification purposes, I am not picking on insurance carriers nor suggesting they are disreputable. As I have already stated, their loss control professionals are simply doing their jobs to protect their clients as well as their employer. However, at times their approach does not necessarily coincide with what is in the best financial interest or operating strategy of your organization. That *is your job as a safety professional, who is employed by your organization to determine what makes the most sense for managing the identified risk.*

The latter scenario is one that will require the use of value-based metrics in the decision-making process.

As an example, one day your boss walks into your office and drops an insurance report on your desk. The report recommends that you upgrade your sprinkler system in the finished goods warehouse to a Highly Protected Risk (HPR) status. The rationale is that, due to the increased height in storage, the value of the finished product and the packaging materials that current sprinkler system does not have the sufficient densities to supply water to control a fire. You need to upgrade your system.

The insurance report does an excellent job of pointing out the required sprinkler densities and the rationale behind their request. Their logic from this standpoint is irrefutable. They are using numbers and logic. You had best do the same.

Using value add metrics, we need to look more deeply into the matter.

The cost to upgrade the warehouse system is $850,000. The value of the product in the warehouse is $2,200,000. The replacement cost for the warehouse is $3,200,000. Based on these numbers, you can start down the path of determining several factors.

1. What is management's appetite for risk?
2. If you do nothing, what are the consequences?
3. What are the hidden costs? What kind of business interruption would the loss of the warehouse cause your organization?
4. What viable alternatives do I have?

Let us examine each of these elements.

Before you can gauge your management's appetite for risk, you have to put together the numbers that provide the worst-case scenario. If the insurance report is correct and the warehouse was a total loss, what is that number including the business interruption.

Furthermore, when dealing with business interruption, it is not always a simple answer of "don't worry we have insurance to cover that." Loss of market share during an interruption can frequently turn into a permanent loss of market share even when you recover. Your customers may find that the generic pill works just as good as yours or the alternative product they switched to is satisfactory. So why switch back? No BI coverage can address that.

The first number (worst-case scenario) is a must have; the second number is a supporting figure that management would have to digest to determine their appetite for risk.

Then, we must ask the insurance carrier what happens if the decision is to leave things as they are. Does this mean the rates will go up? If so, how much? Does it mean the warehouse could be dropped from coverage? If it is the latter case, that is a different matter that raises the stakes.

Then, there are the alternative solutions to be considered. What if you leased additional warehouse storage to lower the level of storage so that the densities come back into insurance carrier acceptable compliance levels? Alternatively, what if you changed packaging materials or rearranged storage?

Even the alternative solutions can require cost analysis. Let us say that instead of a finished product warehouse, we are dealing with a raw materials warehouse. Your solution is to move the product to an alternate location to reduce storage. Just when you believe you have found a solution, you find that you can't because of the new Just in Time (JIT) production process that has been developed. In the new JIT system, raw materials are "locked in" to their locations in the warehouse to assist in the new approach. Now your cost justification must be compared to the one they submitted to justify the JIT approach.

In order to answer all of these questions, the modern safety professional will have to start thinking outside of "regulatory mode" and switch to thinking in terms of dollar signs. The answers to most of the aforementioned questions, as well as many of the solutions boil down to dollars and cents.

As with the other scenarios, many of your answers are going to come in value-based metrics, not incident rates or traditional metrics.

FAILURE TO CHANGE IS NOT AN OPTION

In more than one instance in the previous chapters, I have pointed out that there is no magic formula for what metrics are best for an organization. The answer is normally "it depends," which is the simple truth. Much depends on what the organization values as success.

However, in the case of value-based metrics, the safety professional will either have to learn to adopt them or remain mired in the past of attempting to make business cases without a businesslike approach.

As our profession continues to "professionalize" there will be an increasing need to understand and use value-based metrics. These new metrics will become the main metric of measuring success, justifying capital expenditures, and answering what controls make the most sense as an answer to addressing identified risks.

The phrase "appetite for risk" has worked its way into the business lexicon of all organizations. This phrase will not be quantified by vague measurements. You will not measure this "appetite" with a large, medium, or small approach. The logic to arrive at just how big that appetite is will change for the topic being measured, and that measurement will end up being defined by the dollar metric.

Failure to change to this value-based metric is not an option for the future safety professional. It is a necessity.

REFERENCE

Lopez, C. G. *Managing Risk; Not Safety.*

8 Using Balanced Scorecards

If you don't collect any metrics, you're flying blind. If you collect and focus on too many, they may be obstructing your field of view.

Scott M. Graffius
Agile Scrum: Your Quick Start Guide with Step-by-Step Instructions

If we are to measure on more than one dimension, we will have to report on more than one dimension as well. The simplest means of doing so is a method of reporting that is not peculiar to safety. In fact, we cannot even claim that we invented it. It is known as the "balanced scorecard," or as it is also called the "mixed scorecard."

The concept is simple enough. Find several metrics you consider as reflecting measurements that are viewed as important by management and begin to track them. We will cover what is important to management in the next chapter, so for now let us concentrate or what has been defined as "important" or as we have described several times, the measurements of what represents success.

The concept of what makes any scorecard "balanced," or "mixed" is in the eyes of the beholder. For the sake of consistency, I will use the term "balanced" to represent both the concepts of "balanced" or "mixed" in the following pages. In fact, I am sure some organizations have other terms to describe this approach to metrics. Consequently, let us focus on the approach not the terminology.

Continuing the theme of how metrics will vary from organization to organization, there is not a firm policy or standard on what numbers should be tracked or for that matter how many should be tracked. What the balanced scorecard offers is the ability to look at more than one area of risk to the organization and measure it accordingly.

DESIGNING THE BALANCED SCORECARD

The perfect design for a balanced scorecard is the one that delivers the message of what the organization views as success. As stated previously, this is in the eyes of the beholder regarding what one defines as the perfect set of metrics that measure success. This "perfect" balance is also subject to change from time to time as an organization's priorities change.

However, to paraphrase the quote at the beginning of the chapter, if we are not measuring something we are flying blind, if we are measuring too much, I have destroyed your focus and you cannot see where you are going, or, in some cases, where you have been. This is our first objective. How much is too much or too little? Have I captured the essential metrics or missed something that should have received our focus?

This brings us back full circle to what we report will look different at different levels of the management ladder. Consequently, we should be designing our scorecards understanding this difference in content.

I believe, I have made a sufficient and compelling case for upper management requiring a more concise scorecard with less metrics. This will allow them to keep the focus we seek in their attention to the reported safety statistics, as well as their roles in maintaining the safety culture of their organization.

As we move down the management chain, we will find the scorecards looking much different. They can be (but are not required to be) more expansive in tracking metrics. This especially applies where we want to track leading indicators as part of our reporting system.

In the Preface of this book, I apologized for what I said was going to be a biased view toward metrics considering I had spent my entire career in profit manufacturing. I will temper that apology a bit now by applying the follow-up experience I have gone through with other types of organizations. When I retired from manufacturing and moved to the consulting field, I encountered many of the same issues with metrics I had seen in my previous career path. There was a plethora of lack of focus.

THE FOG OF STATISTICS

One of the first clients I handled in my new role as a consultant asked that we take over the loss control reports they were getting from their current broker. The first time I saw the report, I was astounded. It was 125 pages long. It was a thing of beauty. It had graphs and statistics for every conceivable loss and cause of loss that was being tracked. It further broke all of these down in the same painstaking detail for each of the company's four locations.

My first thought was that this report was prepared for their safety people to review and analyze their losses and decide on what area to target. However, as it turns out, the report was for the executive team to review at their quarterly business meetings.

Although the temptation was to walk into their next business meeting and deliver a diatribe on just how bloated and unfocused their current report structure was, things are not just done that way. Despite the fact that they were a business school study for the data overload, change is better received if it comes slowly.

The report was also a perfect example of what we must avoid as safety professionals. Know your target audience and the message you are trying to deliver. The previous person putting this report together was clearly of the view that more was better. I call this the "volume would impress them" theory of reporting. If nothing else, the pretty charts and graphs would wow them. The concept that volume means quality is a dangerous concept for any of us to fall into. Or worse, that volume demonstrates hard work and thorough work.

In my experience, quality normally wins over quantity when it comes to metrics. Further to this point, the higher you go the more succinct the report the better. Unless you are making some kind of in-depth study or engaged in a research project that needs statistical support, we should be focusing on concise, not a tome.

Consequently, in the interest of not introducing a culture shock of change, we duplicated the initial report in the identical format as they had been receiving it.

However, for the next report, we included the first executive summary at the front of the report.

This particular organization was labor intense so consequently their focus was on workers' compensation costs (and premiums). Their losses and premiums have been steadily rising. They knew enough that they had a problem. They just were not sure what the problem was nor what to do about it. As discussed in the previous chapter, no one had to explain to them that premiums going up each year was not a good thing. That was a metric they understood. Where they were in a quandary was what to do about it.

Looking for an answer in a 125-page report might work for a seasoned safety professional who had the time to devote to the analytics. Expecting an executive team reviewing the document once a quarter to see through the "fog of statistics" was not going to get it done.

The first executive summary to the report included a chart that "normalized" their numbers. Despite all of the various charts and graphs in the 125 pages, in no place had the statistics been normalized for their four locations, other than the LTA rate that was buried in the multitude of charts.

Because it was employee injuries and illnesses, they were focused on the report I developed. In Table 8.1, you will see our first attempt at a crude balanced scorecard that removed some of the fog to help management identify what they wanted to track. Included in this scorecard was a goal to target and a comparative rate that represented how similar organizations were performing.

At this juncture, I would like to pause for a second to answer those that might look at this and ask if it is truly a balanced scorecard. The first misconception we must clear up about a balanced scorecard is that it absolutely requires a mix of leading and lagging indicators. The second is that it should not be just one type of loss you are tracking. Neither of these assumptions is correct.

The balanced scorecard can be considered "balanced" even if it is only one metric you are tracking but in different ways. As can be seen from the afore-mentioned chart, the numbers are expressed in traditional rates for calculating employee injuries and illnesses. Why? In this particular ,organization that was the desired metric. Because the entire 125-page report reflected these values, we could hardly depart from them because it would make the remainder of the report moot.

TABLE 8.1
Multisite Lagging Scorecard

Location	Compensable Rate	Recordable Rate	LTA Rate	NAICS Rate	Goals
Site A	25.3	9.8	7.9	7.1	6.0
Site B	14.8	8.8	8.1	7.1	6.0
Site C	10.9	3.5	2.1	7.1	6.0
Site D	5.5	5.5	5.5	7.1	6.0
Total	14.1	6.9	5.8	7.1	6.0

I realize that, at this point, some readers may be saying "but you told us that this was the Body Count metric problem." I stand hoisted on my own petard. Actually, I stand by my comments in that chapter. They were not meant to say do not ever use that metric, nor that employee safety is not paramount in our mission. My intent was that where applicable we expand our vision of metrics. As you will see, we did just that with this organization from a value-based metric. However, their chief concern of what they wanted to track was just employee injuries and illnesses. That was their focus. I will never argue against upper management having focus.

If it were my company, would I have limited my focus to just that single loss stream? No, but this is what they chose as a business team. The important point was that we presented them with different options. But they chose these as their metrics of choice.

What this brings us back to is the repeated concept of there is no "set" numbers that have to be reported. Are there some better than others? Of course, but that is at the discretion of the upper management of the organization to make that decision. The safety professional provides the options.

As the quarterly meetings progressed, the numbers began to tell a story. The NAICS rate was introduced into the report to allow this organization to see where they stood with comparative organizations in their industry group. The goals were set so that locations could get an understanding of what target they should be aspiring to attain.

Despite the very basic nature of this original balanced scorecard, it was beginning to generate questions at the business meetings. Why was the LTA rate for this organization within such close range of the TRIR? This led to questions about how well their restricted duty program was working or if it was being used at all.

The next leap taken in this journey was to agree to place "value-based metrics" into the executive summary. This led to the following (Table 8.2) being inserted as part of the report.

Soon their management clamored for a combination of the two, which resulted in Table 8.3.

As time marched on the executive management decided that their focus should be on the lack of a restricted work program and using the cost per service hour (CPSH) metric as their severity index. This made the need for tracking the compensability rate secondary. Consequently, they decided to remove it. They were

TABLE 8.2
Cost Breakdown for Grounding and Bonding

Location	Total Reserve	Cost per Man-Hour
Site A	$60,809	14¢
Site B	$169,926	30¢
Site C	$12,631	2¢
Site D	$11,661	4¢
Total	**$255,027**	**14¢**

TABLE 8.3
Multiplant Incident Rates

Location	Compensable Rate	TRIR Rate	LTA Rate	TRIR Goals	Total Reserve	Cost per Service-Hour
Site A	25.3	9.8	7.9	6.0	$60,809	14¢
Site B	14.8	8.8	8.1	6.0	$169,926	30¢
Site C	10.9	3.5	2.1	6.0	$12,631	2¢
Site D	5.5	5.5	5.5	6.0	$11,661	4¢
Total	*14.1*	*6.9*	*5.8*	*6.0*	**$255,027**	*14¢*

evolving. This also demonstrates how balanced scorecards are not carved into granite. They can be changed. Their executive management was focusing.

A year later, their chart looked like Table 8.4.

This case is an example of "management evolution" of how the proper reporting of safety statistics can create a management focus. How working with management is the best way to select the proper metrics to gain this focus. In the period of one year, this evolution was practically a revolution for their executive team. Through trial and error, they had selected the metrics they found as meaningful. The operative word being "selected." For the first time, they were not innocent bystanders, they were involved in the selection process. This process resulted in a shift in thinking of their business team. Or should I say the beginning of a focused approach to managing their safety.

Through this entire exercise, we must not forget that the objective was not simply to come up with a perfect scorecard as a theoretical exercise. It was to come up with one that would impact their main concern which was the problem of rising workers' compensation insurance premiums. It was keeping their eye on this goal that they agreed that the CPSH rate (the value metric) was telling a clearer story than their previous compensability rate number. Consequently, this influenced their decision to quit tracking that particular rate and focus on the CPSH.

They also found that their restricted duty program needed closer tracking so they added a DART rate to compare against their LTA rate. Suddenly, lost time accidents

TABLE 8.4
Multiplant Cost Rates

Location	TRIR Rate	DART Rate	LTA Rate	TRIR Goals	Total Reserve	Cost per Service-Hour
Site A	5.1	3.8	.9	6.0	$15,222	5¢
Site B	4.2	3.8	1.1	6.0	$59,126	20¢
Site C	8.8	6.5	1.1	6.0	$2631	1¢
Site D	2.1	1.5	.5	6.0	$5666	2¢
Total	*5.1*	*3.9*	*.9*	*6.0*	**$82,645**	*5¢*

dropped precipitously as the sites embraced the restricted work programs. I never recommend the tail wagging the dog, but here was a simple exercise in the age old law of we manage what we measure. By creating a metric focused on the restricted duty program, the sites got the message. Get it working, the boss is watching. This also had an obvious ripple effect on the cost of the incidents as well. As multiple insurance studies had demonstrated, getting the worker back to work, even on a restricted duty basis, drops the cost of the loss as opposed to leaving them at home.

This journey of focusing on metrics that measure success is typical for an organization that suddenly realized that you manage what you measure.

As a final note on this particular organization, it was not long before they jettisoned 110 of the 125-page report. They had reached a point where the executive summary was providing the focused information they needed. The average executive summary was now two pages. Supplying charts to illustrate the tables in the executive summary expanded the front of the report to 15 pages. Soon it was decided that the old voluminous format was not necessary. The new report length was 15 pages maximum.

They began to understand that you are flying blind not just with lack of metrics but also by having so many metrics that they block your field of vision.

BALANCED SCORECARDS WITH LEADING INDICATORS

Where the balanced scorecard is of inordinate value is when one is measuring leading indicators. As noted in the chapter on leading indicators, they must have a number and that number must be made normative for sensible reporting of the safety data.

Leading indicators need a report format that mimics to a large extent what we have done for years with lagging indicators. Ambiguous and raw number reports do not tell the tale.

These reporting scorecards can be arranged in a variety of manners with different approaches. I will demonstrate a couple of these approaches using contractors and contractor projects as an example.

The first balanced scorecard we will look at is for a single contractor working on one project. This could certainly be expanded to one contractor on multiple projects as well. The second scorecard we look at will be for multiple contractors working on multiple projects.

For the first example, our single contractor working on one project, we have Table 8.5, which provides a sampling of what can be easily tracked.

This particular balanced scorecard was designed to track contractor activities for an Owner Controlled Insurance Program (OCIP). In this type of insurance arrangement, there is an added incentive to tracking contractor activities because losses would be covered by the project owner's insurance plan.

In order to avoid any confusion, the insurance plan in this case certainly gives incentive for this type of metric tracking, but does not dictate the items tracked on the balanced scorecard. One could just as well select the same metrics for a normal insurance program that is not owner controlled.

In this case, we see that the balanced scorecard takes on more of the traditional look we think of in a balanced scorecard. The losses were not limited to just

TABLE 8.5
Contractor Leading Indicator Scorecard

Item	Criteria	Result
Incident Report Efficiency	Within 24 hours	% of submissions on time
Monthly Safety Meetings	Department rep attending monthly meeting	Attendance
NEO Training	Monthly logs of trained employees	% of new employees trained
Qualified Person List	Monthly submission of new qualified persons	Submission of report
Critical Lift Plan Notification	72 hours in advance	% of submission of plan for all lifts
Dig Request Notification	24 hours in advance for excavations >5 ft.	Submission of dig request
Response Time on Inspection Reports	48 hours from receiving report	% if on time submissions
Incident Rate Submissions	Fifth day monthly	TRIR rate
Loss Rate	$/¢ per man-hour	Cost (adjusted)

employee injuries and illnesses. Property damage and third-party liability concerns are evident as well. Consequently, the focus on Critical Lift Plans, Qualified Persons lists, and Dig Requests are included to track these types of losses. Why these three metrics? Once again we need to select "focus" metrics that are telling us how the risk controls are holding up on the work site. The owner of this plan and project could have selected others, but these three were selected as representing a contractor's adherence to the safety culture that was being promoted for this project.

The Critical Lift Plan addresses all heavy lifts of equipment. More than one accident has occurred due to crane failures due to improper rigging, misuse of equipment, overloading, operator error, and other factors. This type of loss is not restricted to employee injuries. The type of property damage that can result from failed lifts on a construction site can lead to large property losses.

This is also the case with the Dig Requests. This metric addresses *three* concerns in one control system. First, there is the employee safety factor during trenching and excavation operations. Anyone who has spent time doing underground work on a construction site can attest to the fact that without safety considerations in place, a lot can go wrong, which can lead to an extremely serious loss.

Second, without proper knowledge of what is in the ground where one is digging, it does not take much to hit underground utility pipes, fiber optic cables, and electrical lines. To avoid this hazard, there are call in systems and on-site access to "As Built Drawings" to avoid hitting these lines and causing costly third-party damage.

Third, these same dig incidents that damage pipes, fiber optics, and electrical lines can lead to property loss as well.

Finally, the Qualified Persons list is intended to cover both of the aforementioned issues as well as other hazardous operations that require permit sign offs, pre-job evaluations, and operational checks. The Qualified Person list would indicate who is, for lack of a better term, "qualified" to conduct the proper risk assessments and inspections that allow the work to proceed.

Careful study of these leading indicators makes it evident that they were not randomly selected. In the chapter on leading indicators, this matter of selection was discussed as being paramount to the success of leading indicators. Furthermore, these selections pass the Practicality and Sensibility Tests, I emphasized in earlier chapters. All of the data to build these reports is readily available and more importantly has been approved by management to represent what they consider important in measuring success. Each was selected based on how they represented a portion of the safety program that the owner of the project felt needed close monitoring.

As a final note on this list of leading indicators, I will clarify a point I am often asked about leading indicator measurements. If we look at the third column, one could speculate that these are all lagging indicators. As we covered in Chapter 5, addressing leading indicators, there must be a measurement on everything for it to be a metric. For it to be a good metric, it must be a number and for that number to make sense it must be put into a normative format. Consequently, all indicators whether they are leading, lagging, or value based must have a metric attached to them, which we could consider results based, or lagging in nature. This does not take away from the "action" of the leading indicator, which is the objective of the exercise. It is simply meant to measure the success of the effort.

MEASURING PERFORMANCE ON MULTICONTRACTOR SITES

As I stated previously, the second example of a scorecard we will look at is the multiple contractor scorecard. This normally means not only multiple contractors but also multiple projects that require being measured at the same time.

One of the great challenges any organization has is to manage the control of multiple contractors on a given work site, or worse on multiple worksites at the same time. This especially becomes the case in two scenarios.

Scenario 1

If they are working on a site where their actions could impact your operations in a negative way, more stringent controls are needed to coordinate their work that activates with your production activities. These are normally classified as "Greenfield" and "Brownfield" projects. Greenfield meaning that they are completely detached from your operations building on a site that is either in another location and cannot impact your operations. Brownfield can mean they are either building adjacent to your existing operations with the ability to impact you ranging from remote to a daily concern.

As an example, when contractors worked on Greenfield projects for us, we would give them safety rules to follow but would for the most part put them in charge of what was happening on their construction site. After all that was why we hired them. For Brownfield, the situation changed dramatically. In our chemical and pharma plants, they were closely managed to ensure that they followed our contractor safety rules. This management including the best we could do to isolate their operations and strict daily communications to determine if what they were doing could impact our operations. They did not break a line, enter a confined space, lock out a system to name a few activities without getting a signed off Safe Work Permit from our management.

Scenario 2

The second scenario is the aforementioned OCIPs, or even its close relative Contractor Insurance Programs (CIPs). In both instances, the "owner" has a massive incentive to ensure safety in operations are being run as expected.

In either instance, the use of a balanced scorecard can (like the previous example) provide a report on how safety compliance and resultant losses are occurring from month to month. They also serve as a reminder to the contractor that their performance is being closely monitored.

As a note, these same approaches can be applied to non-contractor multi-site locations but for this example we will focus on contractor operations.

The challenge once again becomes how much information can you track before you lose focus. In this example, this "recipe" is now made more complex because you are not tracking a single contractor but multiple contractors and you are doing it on multiple projects.

The scorecard (Table 8.6) is an illustration of how this scorecard can be designed to measure these multiple contractors on multiple projects. In this example, the metric choices have been limited to three metrics that have been identified as the key markers to success.

In an effort to not fly blind and to maintain focus, the "owner" of all of these projects decided that all the contractors working on the various projects going on simultaneously would be measured by the same metrics.

The three metrics that have been selected are:

1. CIP Safety Program Compliance
2. Incident Rate (TRIR)
3. Workers' Compensation Loss Rate

Of these three metrics, the most problematic to assess would be CIP Safety Program Compliance. Because this metric is not as definitive as the two lagging indicators, some clarity will be necessary to define how the measurement is arrived at for this portion of the metric.

The solution arrived at for this particular measurement was to identify the program elements that represent compliance with the CIP Safety Program. These are:

1. Incident Report Efficiency
2. Monthly CIP Safety Meeting Attendance

3. New Employee Orientation Training
4. Response Time to On-Site Inspection Reports
5. Timeliness of Pre-Notification for Critical Activities

TABLE 8.6
Project Summary Scorecard

Project Scorecard Summary (Year-To-Date)

Project	General Contractor	Items	Results
Terminal One Modernization Project	CGL Construction	CIP Program Compliance	80%
		Incident Rate	3.57
		Loss Rate	$.50 per man-hour
East Expansion Project	ABC Construction	CIP Program Compliance	79%
		Incident Rate	0.64
		Loss Rate	$1.00 per man-hour
West Gate Replacement	Ajax Construction	CIP Program Compliance	92%
		Incident Rate	10.26
		Loss Rate	$4.76 per man-hour
Digest Cover Replacement	Achilles Construction	CIP Program Compliance	70%
		Incident Rate	0.00
		Loss Rate	$0 per man-hour
Generator 4 Installation	Miller Electric Service	CIP Program Compliance	53%
		Incident Rate	0.00
		Loss Rate	$.76 per man-hour
Injection Well Booster Pump	Hector Construction	CIP Program Compliance	43%
		Incident Rate	0.00
		Loss Rate	$.10 per man-hour
Aircraft Parking Apron	Paris Construction	CIP Program Compliance	66%
		Incident Rate	12.40
		Loss Rate	$4.96 per man-hour
Water System Improvement	Ocean Bay Construction	CIP Program Compliance	77%
		Incident Rate	0.00
		Loss Rate	$1.09 per man-hour
New Courthouse	Priam Construction	CIP Program Compliance	23%
		Incident Rate	0.00
		Loss Rate	$0 per man-hour
Paradise Road Expansion	Marx Brothers Construction	CIP Program Compliance	61%
		Incident Rate	0.00
		Loss Rate	$1.12 per man-hour
Pedestrian Bridges	L & H Construction	CIP Program Compliance	14%
		Incident Rate	0.00
		Loss Rate	$0 per man-hour
Terrazzo Floors Installation	Flooring R Us	CIP Program Compliance	15%
		Incident Rate	0.00
		Loss Rate	$.05 per man-hour

TABLE 8.7
Weighing Table

Item	Points Possible	Weight (%)
Incident Report Efficiency	10	26
Monthly CIP Safety Meeting Attendance	3	8.5
New Employee Orientation Training	5	13.5
Response Time on Inspection Reports	10	26
Pre-Notification of Critical Activities	10	26

One can quickly see that these are all leading indicators, very similar to the ones pointed out in Table 8.5. However, now they have been combined in one measure under the heading of CIP Program Compliance.

Why these five? As with the earlier example, they were the five that the "owner" considered indicative of a contractor complying with their CIP construction safety program. Could there be more? Yes. However, as addressed at the very beginning, the tracking of too many metrics can have a negative effect on the focus of what you are attempting to measure. All five also scored high on the Practicality and Sensibility Test scale.

Could there be different metrics selected? Once again, the answer is yes. However, these particular metrics had a secondary pupose. They were selected not only for CIP program compliance but also so that they owner could "grade" the contractors on the project for future bidding purposes. Those who scored well would go to the front of the line for future bids. Those that did not score well might be eliminated from future bid lists.

This latter feature lends yet another mission to providing good metrics on a balanced scorecard that can have a ripple effect for the contractors.

WEIGHING SCORECARDS

As we look at the five indicators selected to measure compliance with the CIP Safety Program, one could argue that some carry more difficulty and effort than the others. This is easily resolved by "weighing" the various subcomponents. Table 8.7 demonstrates how this "weighing" of the various subcomponents can be easily accomplished.

Weighing of components in evaluation lists is not a novel concept. But it does add validity to the concern that simply attending a monthly safety meeting does not compare with response time to inspection reports. Although not necessary, weighing adds an additional dimension to the balanced scorecard, especially when one of the three indicators is complex.

The weighing formula in the aforementioned table is normally conceived by the safety professional and offered as a further definition of safety performance to the owner.

As for the Practicality Test concern, in terms of collecting and calculating this data, with modern Excel spread sheets, the raw data can easily be inserted into a spreadsheet, a formula put into the sheet, and the calculation spit out by the spreadsheet.

Table 8.6 also illustrates the mix of leading and lagging indicators. Of the three metrics we have:

1. Leading Indicator: Compliance with the safety program the contractor has committed to follow.
2. Lagging Indicator: Use of the traditional TRIR.
3. Lagging Indicator (value metric): Cost per man-hour for workers' compensation losses on the project.

Although there is no magic rule that says one must only pick these three metrics, as noted in this particular case, the owner wanted to keep the number of metrics tracked to a simple number.

TIMING OF BALANCED SCORECARD REPORTS

The timing of how often to submit balanced scorecards depends greatly on what you are reporting as well as for whom your report is designed. Most executives want to see reports on a monthly basis. However, this is no hard and fast rule. In the example just given for the construction project, the owner was satisfied with a quarterly report.

For the most part, reporting frequency for safety reports often coincides with the frequency a manager that is responsible for achieving the metric sees other similar reports. If a plant manager is used to seeing monthly production reports, they will most likely ask to see monthly safety reports.

Frequency can also be dictated by timeliness of data. For those reports that contain workers' comp cost information, one rationale for a quarterly basis, as opposed to monthly, is that the cost information does not change that significantly from month to month.

Whether you are issuing the report on a monthly or quarterly basis, one of the considerations must be if the costs are adjusted to reflect the true "mature value" of the claim. This was covered in an earlier chapter about reliable data. We must make certain that the data reflect the true cost. If it is not adjusted, it is upon the person compiling the report to make note of the fact that the cost will most likely change in the future.

The final answer to timing of reports is that it depends on how often management wants to see them.

MULTIPLE METRIC REPORTING

In the case of the earlier example of the multicontractor multiproject report, we discussed the three metrics that were selected and why. However, a balanced scorecard can include several metrics for those who want more to track. Table 8.8 gives an illustration of tracking multiple metrics on a scorecard. In this case, the scorecard captures the following information:

1. Reportable Rate includes all cases including first aid and near misses
2. Compensable Rate includes all cases that are workers' comp compensable

3. OSHA TRIR includes traditional OSHA recordables
4. LTA Rate includes all traditional lost time accidents
5. Workers' Comp Costs includes workers' comp reserves (unadjusted for maturing)

Now there are five metrics that management which is tracking. All of them are lagging indicator rates. All of them are targeting only employee injuries and ill-nesses. In this organization's case, they had a fleet but were not tracking fleet losses unless a person was injured in a fleet accident and claimed workers' compensation. They were also ignoring third-party General Liability (GL) cases.

Unlike the prior example of the organization that was labor intense and had no other significant losses, it choose to measure, in this case this type of balanced scorecard could just barely meet the definition of "balanced" for their risk picture. Despite doing an excellent job of measuring in several different areas, they have only focused on injuries and illnesses.

The second weakness of a card such as this is that since the five metrics they are tracking are all employee injuries and illnesses, which one is a priority measure? This is yet another example of how to many metrics blur the focus.

TABLE 8.8
Multisite Spreadsheet

Widget International Manufacturing
Loss Statistics
Q4-YE

Location	TRIR Rate	DART Rate	LTA Rate	Fleet Cost PMD	Prop Damage	Incident Number	LTA Rate	LTA #	Workers Comp. Costs		E
Manufacturing Location											
Pittsburgh	23.3	9.8	7.9	50	21	17	0.5	1	$ 0.14	$ 60,809	
miami	14.8	8.8	8.1	42	25	23	1.8	5	$ 0.30	$ 169,926	
Los Angeles	10.9	3.5	2.1	31	10	6	2.1	6	$ 0.02	$ 12,631	
Phoenix	5.5	5.5	5.5	8	8	8	2.1	3	$ 0.04	$ 11,661	
Subtotal	14.1	6.9	5.8	131	64	54	1.6	15	$ 0.14	$ 255,027	
Distribution											
Tucson	7.1	6.5	5.8	27	25	22	3.1	12	$ 0.10	$ 73,885	
Houston	1.5	1.5	0.8	2	2	1	0.8	1	$ 0.01	$ 3,463	
Wheeling	6.3	5.6	2.1	9	8	3	0.7	1	$ 0.13	$ 36,016	
Ft. Lauderdale	2.5	2.5	2.5	1	1	1	2.5	1	$ 0.01	$ 458	
Charlotte	0.0	0.0	0.0	0	0	0	0.0	0	$ -	$ -	
Tampa	8.1	5.8	1.2	7	5	1	2.3	2	0.03	$ 4,521	
Subtotal	5.7	5.1	3.5	46	41	28	1.7	14	$ 0.07	$ 118,343	
Totals	10.2	6.1	4.7	177	105	82	1.7	29	$ 0.11	$ 373,370	

TABLE 8.9
Multisite Mixed Cost Spreadsheet

Location	TRIR Rate	DART Rate	LTA Rate	Fleet Cost PMD	Prop Damage	GL Number	Workers Comp. Costs	
Manufacturing Locations								
Pittsburgh	13.3	9.8	7.9	$7	$80,000	$0	$ 0.14	$ 60,809
miami	4.8	1.8	.1	$12	$25,000	$2300	$ 0.30	$ 169,926
Los Angeles	2.9	.5	.1	$1	$300,000	$6500	$ 0.02	$ 12,631
Phoenix	5.5	2.5	.5	$2	$69,999	$129,999	$ 0.04	$ 11,661
Subtotal	6.1	3.3	2.1	5.1	474,999	138,799	$ 0.14	$ 255,027

At the department or plant level this many metrics tracking the same topic does not pose that great of a dilemma as long as the management is agreed on what the priority metrics are. At the upper management level it becomes problematic.

TABLE 8.10
Quarterly Multi-Incident Tracking Histogram

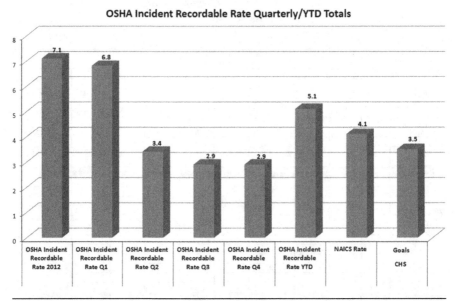
OSHA Incident Recordable Rate Quarterly/YTD Totals

TABLE 8.11

Quarterly Multi-Incident Tracking Histogram

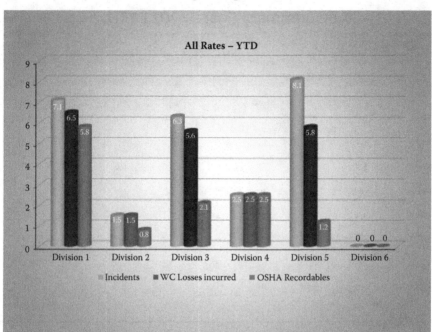

A better approach at tracking multiple metrics is shown in Table 8.9, which captures the following:

1. TRIR
2. DART
3. LTA
4. Fleet Cost per Mile Property Damage
5. GL Loss
6. Workers' Compensation Loss (adjusted)

This gives the balanced scorecard much more balance in terms of tracking metrics in different areas of risk to the organization.

As with all balanced scorecards, Table 8.9 identifies metrics that reflect the risks that the organization is exposed to and agreed upon metrics to measure these risks.

The sole flaw with this table is that some of these numbers are expressed in raw numbers, such as the GL column that expresses what the numbers were for claims. This also applies for the property damage column. In this particular example, that is how the management wanted these values expressed. However, either of these metrics could be normalized if the organization desired so. For example, the GL

TABLE 8.12

Multidivision Summary Chart

ABC Summary Chart – 2021 YTD

Location	OSHA Incident Recordable Rate 2013	OSHA Incident Recordable Rate Q1	OSHA Incident Recordable Rate Q2	OSHA Incident Recordable Rate Q3	OSHA Incident Recordable Rate Q4	OSHA Incident Recordable Rate YTD	NAICS Rate	Corporate Goals
Site 1	4.5	3.9	4.1	3.2	3.8	3.9	4.1	3.5
Site 2	2.2	0	1.9	0	2.0	3.1	2.0	3.5
Site 3	6.4	0	5.2	6.3	5.8	5.7	3.7	3.5
Site 4	5.9	0	5.8	6.1	6.0	6.0	2.7	3.5
Site 5	0	0	0	0	0	0	3.3	3.5
Site 6	4.4	2.8	1.6	2.0	3.1	2.8	8.1	5.0
Site 7	6.5	6.5	6.0	6.7	6.2	6.5	8.1	5.0
Site 8	3.6	0	4.1	0	3.2	2.9	8.1	5.0
Site 9	1.9	2.4	1.5	1.0	3.8	2.4	8.1	5.0

could be normalized to $ per sale, or perhaps $ per man-hour. The selection of the normalizing formula would depend on where the GL claims were coming from and value-based metric formats used in other areas of the organization.

Property losses could also be normalized using $ per the assigned value on property schedules or $ per plant size, or perhaps hazard class. Once again, the first place the safety professional should go to find a formula that will resonate with management is to see what they are using in other areas of the organization to normalize numbers.

Almost any number can be normalized if the organization desires so.

Even with the ones normalized, the normalizations they select could be changed, but the topics they have selected will be maintained. As an example, perhaps the "rate" they choose for fleet cost is not based on a "per mile driven" number. It could be on 100 miles driven, or 1000 miles driven, or cost per vehicle to operate.

USE OF GRAPHICS TO PROVIDE METRICS

Most of the illustrations I have used to demonstrate different approaches to metrics have been tables with numbers. However, this is not to meant to imply that other more graphic representations are also not available. Table 8.10 shows the use of a histogram to illustrate the OSHA TRIR rates of an organization.

TABLE 8.13
Dashboard Example

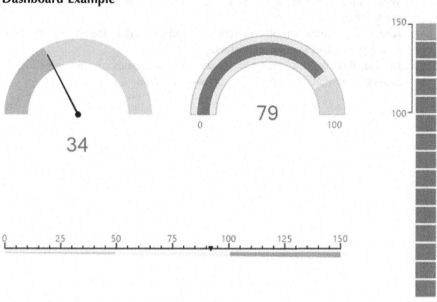

The histogram in Table 8.11 does much the same but, in this case, it illustrates three values in one histogram. Once again, the graphics used are in the eyes of the beholder. What we as safety professionals must not get caught up in is that the graphic is the show and not the metric.

Table 8.12 shows a simple representation of the numbers listed and tells a story as clearly as any pie chart or histogram. The metrics are the focus, not the presentation.

Perhaps of all the graphics used to demonstrate balanced scorecards, none is more popular than the dashboard example displayed in Table 8.13.

Using the "level" indicators that resemble the fuel gauges on cars, we get an immediate picture of where each of the metrics stands in a single glance against the goal or as a score.

FINAL TENETS FOR THE BALANCED SCORECARD

Probably, no more powerful tool exists for the modern safety professional than the balanced scorecard. Consequently, I will leave the reader with some final tenets on using these scorecards.

1. Keep the number of metrics on the card focused to the level of management you are addressing.
2. Select metrics that impact the goals you are trying to achieve and represent the success of the safety efforts.

3. Select metrics that meet the Practicality Test to gather and Sensibility Test to promote.
4. Weigh metrics when including multiple measurements under one metric on the scorecard.
5. Allow for expansion of the scorecard when addressing lower levels of management and leading indicators.
6. Remain flexible and change scorecards when metrics are not targeting the goals of eliminating, controlling, or reducing risk.

9 Selling Upper Management on Metrics That Define Success

We must constantly examine our roles and what value we bring to the organization, then we must be able to demonstrate that value.

C. Gary Lopez

You've got to tell a story, paint a vision, know your metrics and sell, sell, sell.

Mitch Harper

As an avid reader, from time to time, I pick up books on management approaches that have been used by different CEOs. I suppose I got started down this path when I was required to read Joseph Heller's novel *Catch 22*. The novel was a parody of the military that can easily be transferred to the corporate world. In fact, it was required reading because of a business course I was taking in college.

For those of you not familiar with the novel, the hero, or should I say antihero, is a bombardier named John Yossarian. Yossarian has had enough of the war and just wants out. So he decides to tell the army air corps flight surgeon that he is crazy. In the army, this is known as a Section 8. However, Yossarian keeps running up against the same problem. As the flight surgeon explains to him, "If you are sane enough to know you are crazy, I obviously cannot classify you as insane." That ladies and gentlemen is a Catch 22.

On more than one occasion, I have felt I was in a Catch 22 trying to sell management on a different metric than the traditional incident rates we have used in the past. Especially since my predecessors had spent years selling them on that very rate.

Years ago, Dan Petersen suggested I read a book by Max DePree, *Leadership Is an Art*. In his book, DePree suggests that a CEO's first obligation is to define reality. He goes on to put this definition in three stages.

1. Reality: Defining what we are today
2. Vision: Defining where we would like to be
3. Process: Defining how to get there

In order to define what we are today, some sort of metrics are in order.

Perhaps no area of safety management reflects a greater disconnect to our mission than our inability to communicate with upper management regarding what metrics are the best to measure safety success in their organization.

115

In an earlier chapter on leading indicators, I gave an example of a safety director who was measuring success in one manner and an executive who was measuring it in another. Such "disconnects" are not infrequent and become problematic when safety is not in lockstep with the rest of the organization's values. Which begs the question, "Why would safety have one set of values and upper management another?"

We can look far and wide for who the guilty parties are for this disconnect, but, I am afraid, the usual suspects will keep turning up. Who are they? We have to look no further for these guilty parties than the nearest mirror.

This is not an indictment that our profession is being lazy about this issue. It is more a matter of being tuned to a different station. Nevertheless, we cannot use lethargy or "tuned out" as an excuse. We have to realize there is a problem and we have to do something about it.

The root problem of this lethargy has been the eternal dependence on the "Body Count" rates as our method of measurement. By this point, I believe I have sufficiently covered more than an ample number of alternatives to supplement these one-dimensional numbers. However, before we can do so, we must learn to "re-sell" our metrics of choice.

I call this a "sale" because that is exactly what it is. We have been telling executives for nearly a century that these are the magic numbers. Now all of a sudden we are switching gears on them. This is not to say that the switching of gears is not necessary, but it will take some rationale of why expanding metrics is a good idea.

Before we make our "sale," we had better know our market. This is not a one-way street. We need to understand what upper management's "hot buttons" are for measuring success. What will appeal to them. In some instances, they will not really know. This will be your job as the safety professional to present them with options.

While you may believe this should be self-evident, it is anything but self-evident. The first reality check we must grasp is that your average CEO, or for that matter any upper manager, is not sitting around at night pondering what safety metrics they covet to measure success in their organizations. In fact, since we have sold them on our body count metrics long ago, they have just accepted it as gospel at this point.

It is the classic "chicken and the egg" conundrum. We sold them on the metric and have now become slaves to it. Breaking the cycle will take selling them new metrics as a definition of success. Which brings us to what CEO's think of in terms of success in our field.

WHAT UPPER MANAGEMENT THINKS

At this juncture, I would love to give you the new "magic" formula and tell you to go forth and do good. Unfortunately, there is no magic formula. What one CEO feels about "success" in a safety program could be entirely different from how another feels. Certainly, some organizations have a more compelling reason to focus on safety because of the nature of their business. If I am conducting underground mining, or construction, or steel manufacturing, safety is high on my list of concerns due to the nature of my operations. If I am in a light manufacturing or even service industry, it may not be as high a prioroty or a concern. That is not to say it is not a concern, but it is not as evident.

The good news for us is that any CEO worth their weight in salt is going to support having a safe organization. Unless they have been living in a cave for the past 50 years, no CEO wants their organization to be viewed as cavalier about safety.

However, CEOs have a lot of responsibilities toward their organization. Typically, most of these responsibilities come with clear metrics that they can see that measure success. Our job is to understand what makes your CEO tick and what they value as metrics that define success.

In an earlier chapter, I promoted the concept of using dollars as a "value metric" instead of abstract rates. While this can be of great value (as I will demonstrate later in this chapter), there may be some CEOs who find applying the dollar to safety as distasteful (I have yet to meet one, but I am told they are out there).

CEO's and the company culture they choose to impart to the public, the stockholders, customers, and investors can vary.

Let's consider some of the following quotes:

> Back in 1920, W.C. Durant, the President of General Motors, stated: *The economic waste resulting from carelessness is appalling, but anyone who stops for a moment to consider the sorrow and desolation which is brought into thousands of lives each year by utter thoughtlessness must feel a new resolve to make a habit of Safety First.*

As one can see from this quote, the idea of Safety First was born with CEOs in the auto industry as early as the 1920s. Now, while this made for a fine quote, I sincerely doubt that Durant's assembly plants could survive a modern OSHA inspection without hundreds of citations for Point of Operation hazards alone. Nevertheless, here was a CEO who realized he had a problem with losses occurring because of accidents in the workplace. He was making a sincere effort to communicate that he was concerned and cared for his workers.

If we fast forward to the future, we have the following quote from the late Paul O'Neill, the CEO of ALCOA (among many others he made about safety) when asked about applying return of investment into spending on safety.

> *I will fire the first accountant that asks for an ROI on a safety investment.*

Paul O'Neill had purposely selected the safety performance of his company as a motivator to get his management team to think "out of the box" not just on safety, but on all topics as he strove to create a new ALCOA.

No less a safety mentor than the late Dan Petersen echoed this message of not mixing dollars, or value-added metrics, with safety. He stated the following message on more than one occasion:

> *It is a very bad message to say safety is about dollars. The real message is; we are doing these things because we care about you, the company cares about you. Don't even talk to me about doing safety for money.*

To put Petersen's comments on this into context, he was focused on employee safety in these comments and was not dwelling on other areas of loss we have discussed such as fleet and property.

Clearly, these CEOs do not care to have safety broken down to a dollar equation. However, there are also management gurus who remind us that upper management has other obligations as well.

Of all of these management gurus, perhaps Peter Drucker is the most respected and certainly one of the most quoted in the top MBA programs. He makes it clear that one of the first orders of business is fiscal responsibility. A logic that is irrefutable when one considers you can have the best safety program in the world, but, if your company is not profitable, it will not be around long, which, in turn, makes that safety program a moot point. His quote on business is as follows:

> *Economic performance is the first responsibility of business ... without it, a business cannot discharge any other responsibilities, cannot be a good employer, a good citizen, a good neighbor.*

Then, we have Michael Muckian, who, in his book *The Complete Idiot's Guide to Finance and Accounting*, states a similar case regarding profitability as a main CEO responsibility:

> *It doesn't matter what your professional responsibilities are, your management responsibilities contribute directly to your company's profitability.*

As we look at these various quotes and points of view, we can endlessly debate the question of "Who is right?" The answer to that question? They all are. We must view "right" as what the organization views as important and the CEO is the leader of that organization.

This does not make our job impossible in determining metrics. It just gives us an additional perspective on which metrics are going to work the best for our organization.

THE MODERN SAFETY PROFESSIONAL AND SELLING METRICS

Considering the aforementioned quotes, which reflect the different thinking of what represents "success," as a profession, we are left with no alternative but to figure out what metrics make sense in our respective organizations. Before we make a "sale," we have to know our market.

What KPIs the organization has set and can we affect these by what we do? As pointed out in the value metrics chapter, you can affect the insurance costs to your organization by managing those things that impact those costs more closely.

As we have seen from the quotes of the various CEOs, they have responsibilities that range from safety to profitability to their stockholders and employees. Why would we want to be married to a single metric? Furthermore, the nature of your organization may make a metric like the "Body Count" almost irrelevant. Your organization risks may lie in other areas.

As an example, if I am running an organization that is a distribution company and my fleet operations are central to my business operations, employee injuries may be the least of my problems. Our job as safety professionals is to identify what the risks are and creating a metric that management agrees with to identify we are being successful or not with our efforts.

In this particular case, I would focus on fleet losses. Both my leading indicators and my lagging indicators should represent what we are doing to prevent them. This means measuring what controls we have in place. Subsequent to this should be our ability to measure how successful these controls are working. In this scenario, despite Mr. Petersen's comments about money not being a factor, I have personally seen the immediate effect that converting to value-based metrics had on any organization that deals with fleet issues. While there are certainly employee injuries to consider under my workers' compensation, I have fleet accident costs and perhaps even worse the general liability costs of hitting another vehicle. Of the three risks areas the last is the most potentially damaging.

On the other hand, as I pointed out in an earlier example, the labor-intense organization that wanted to just see traditional injury rates in the beginning were focused on employee injuries only. However, as time progressed, once they were introduced to value-based metrics, they gravitated toward these. Your "market" is essentially your risks that need to be addressed.

Now that you have defined your "market," one would hope that the upper management of the company would agree with your conclusions of what these risks are and how they could impact your organization. With that accomplished, you have set the stage for your sale.

Your next step in your sale is concluding what metrics management would like to see that represent success and provide options for them to track these measures.

This is where the balanced scorecard comes into play. You may find that upper management wants to see just three metrics, or five, or seven. However, they prefer all of these to be lagging indicators in numbers since they are familiar with seeing other business results in this format. It is at this time that you should also introduce the concept of value-based metrics. One of the easiest sales to make on upper management with metrics is when you switch to the dollar sign. As I have pointed out, they do not require any special training to relate to that metric system.

On the other hand, while the sale of just a few strategic metrics works for upper management, your "sale" to middle management and frontline management may revolve around more metrics including leading indicators.

Naturally, this would give a different "look" for their scorecard. However, all of the metrics you select must complement each other. As discussed in previous chapters, we must ensure that the sale of your leading indicators are targeted at making sure the top-three lagging indicators given to upper management are supported. The bottom tactical metrics should be in place to support the more strategic metrics that upper management is reviewing.

As an example, Tables 9.1 and 9.2 provide metrics that might be used for a distribution company.

TABLE 9.1

Upper Management Scorecard

Category	Metric	Metric Type
Fleet Losses	$ Cost per mile driven	Lagging
Third-Party Impact	$ Cost per third-party incident	Lagging
Employee Injuries	LTA rate for employees injured	Lagging

TABLE 9.2

Middle and Frontline Management Scorecard

Category	Metric	Metric Type
Fleet Losses	$ cost per mile driven	Lagging
Fleet Losses	$ cost of third-party injuries and damages	Lagging
Fleet Training	% of drivers trained on defensive driving	Leading
Fleet Telematics	% of drivers speeding in geo–fenced areas	Leading
Fleet Telematics	% of drivers doing "hard stops"	Leading
Fleet Maintenance Rate	$ cost of fleet vehicle per miles driven	Lagging
Fleet Maintenance	% of vehicles maintained per schedule	Leading
Employee Injuries	LTA Rate	Lagging
Employee Injuries	TRIR Rate	Lagging
Employee Injuries	$ cost of injuries per man-hour	Lagging
Employee Injuries	% of employees who finished ergonomic training	Leading
Employee Injuries	% of employees who utilize load-assisting equipment	Leading

As we can see in these two examples, the upper management scorecards are brief and exclusively lagging-type indicators. There are also only three metrics they have chosen to measure success, or, conversely, failure. The operative words in the aforementioned sentence being "metrics they have chosen."

This would not be untypical of an upper management approach toward measuring anything. As we have already discussed, they have more than safety success on their minds when they are measuring other aspects of the organization. Consequently, they will, in many instances, ask for a quick picture of what indicates the organization is trending in the right direction. It becomes incumbent upon the safety professional to assist them in making these selections. Our role is not to pick as much as to provide options and make a "sale" for the best ones.

If one were speculating on why these three were selected, we would go back to the question of what does the organization do? Since distribution is its key function that means that the methods of storing and distributing the product is the business model. We must look for risks within this model. We must study where has the organization been hurt in the past to search out our metric of choice to sell.

There is an old adage that goes "nothing is better for the safety business then a good old fashioned disaster; as long as it happens to someone else." Since we are a profession that measures much of what we do by loss, there is much truth in this saying. Consequently, we should not limit our study of loss to our organization but also to what is happening in similar organizations and business models.

As an example, after the tank farm fire discussed in an earlier chapter, is there any safety professional in the country (or perhaps the world) who is not out looking at the safety controls in their tank farms? More to the point, how hard is the "sell" going to be for those that discover a lack of controls and recommend improvements?

But the key to the "sale" is to use the value-based dollar costs of that tragedy to extrapolate what the impact would be of a similar loss to your organization.

At the end of the day, these three metrics may not be the all-defining "magic bullet" that puts the appropriate management oversight on all the risks of the organization. However, as with the contractor scorecard in the previous chapter, these should represent key indicators of the safety program trending in the right direction. They also represent a joint effort of the safety professional and upper management to address what they view as the serious risks and to put a number on measuring the success of the controls for these risks. Which brings us to the next scorecard.

As we move to the Table 9.2 scorecard that identifies what middle and frontline management will measure, we see that we have greatly expanded the number of items being measured. We have also departed from a strictly lagging indicator metric system to one that embodies both lagging and leading indicators.

On this second scorecard, we have now addressed all of the three main metrics upper management is looking at, sometimes using the exact metric. However, now we are supporting that metric looking at other areas that need to be measured.

Since we have identified fleet loss as a main target of measure, we are not just mimicking the same measure at a lower level; we are measuring what management controls we have decided to put in place to control this loss. These are the leading indicators that expand the metrics to impact the lagging metrics we know will be the outcome of our successes with the activities we choose to reduce the losses or eliminate them.

All that remains for the sale to middle management is an explanation of how these metrics have been endorsed by upper management and how they can now be measured by the activities expected of them to attain these strategic goals.

This brings us full circle back to our earlier chapter where we discussed leading indicators. The strength of leading indicators is that management has control of its destiny. You have now given them the "things" they can do to meet their goals. What could make a sale easier?

We must also remember that nothing could be a better example of why the "sell" of metrics is a never-ending process with upper, middle, or frontline management than if we did suffer failure as a result of selecting the wrong lagging or leading indicators. As cited in the first fleet example of the division president who was using one fleet metric and the safety director using another, we at least want to be on the same page with agreed-upon metrics. If they fail, there is no shame in going back and changing them. The shame is in thinking they are carved in stone.

In the new world we are creating with metrics that are more diverse, let us say that we missed the mark completely with the leading indicators we selected. A scenario that is not that farfetched. What do we do? Act like nothing has happened? Tell upper management, "Oops sorry about that"? Sulk away and seek employment elsewhere because you missed the mark on your first attempt?

Those are certainly some approaches we could take, but perhaps the more businesslike approach would be to take an objective view of what didn't work and go back to the drawing board to see what does work. This will not make you look weak and indecisive to upper management. It will make us look professional.

They realize that we are safety professionals not seers with a crystal ball. We are making the best assumptions we can based on the knowledge we have at the time. Experience gains us knowledge. If our experiences show that a particular tactic is not working, then you switch to another. In the military, this is a given in tactics and leadership. We need to take the same approach as safety professionals.

SELLING THE SYSTEMS SAFETY APPROACH

One of the biggest "sales" I ever made in my career was when I was put in charge of the EHS and Risk Management of a major IBU and had to decide how to manage the safety of our wide-flung operations.

When installing such management approaches, we are supposed to sell upper management on said approach. Quite frankly, I can't imagine attempting to do this without some type of metrics to support the sale.

When I built the first "systems" management approach in my company back in the late 1980s, the concept had to be "sold" to upper management. At the time, there was no standard to use as a guideline. There was no rationale that we had to do it to gain "certification" of our system. We had to do it for the oldest of reasons. It was good for the business.

I had 47 locations scattered across the globe with a "mother ship" in the United Kingdom stating we would be "World Class" in safety. The world headquarters was issuing "EHS guidelines" of how we were going to get there. Guidelines with which every location, no matter where, would comply. In the United Kingdom, the term "guideline" is the same as "policy" in the United States. I was clear what I had to do. The guidelines were well written. They specified "what" we had to do and left the "how to's" up to us. However, with 47 locations spread across the globe, that was a lot of how to's. Consequently, the systems approach in my case was not so much innovating the latest cool program; it was a matter of management survival. I needed implementation, organization, and consistency across the globe.

That there was none of the various management systems standards, we have now probably acted as the good news and bad news. The bad news was that I had nothing to use as guidance. However, the good news was that I was not tempted to use "certification" as a rationale, and we consequently built our system along the lines of what made sense and learned what to do and not to do from our successes and failures.

The effort to get the system up and running was considerable. My first objective was to sell all of the plant managers of these scattered locations on why we were doing this. I could have fallen on the "corporate says we have to do it argument," but I wanted them engaged. I had to sell them using metrics. Like all practical plant managers, the prevalent question was? "What do I get for this?" In other words, how was I going to measure a benefit for them?

I had benefited from an earlier project I had been asssigned to determine if we should become OSHA VPP sites or not. After looking at the pros and cons, I decided that it simply was not worth the aggravation. Compliance was hardly a problem. We were already doing all the things that a VPP site had to do and more. It was a simple matter of all we got for it was flying a flag and avoiding some general inspections. Neither of these was of much value to us. During this process, I had used dollars to demonstrate what it would cost and what we got in return. As I was later to find out, this approach paid dividends. I had already won over some plant managers as taking a business-savvy approach, instead of simply waving the safety flag. I had established some creability by speaking their language.

For this round, I began to crunch the numbers of what some costly incidents within our company have cost us including a recent tanker unloading incident and an explosion at one of our plants. I paired these numbers into areas we would address as part of our new systems approach. I then added in predicted reduction in dollar cost to the insurance programs. The icing on the cake was me explaing how not meeting the corporate guidelines meant potentially being beaten up during an internal corporate audit the results of which would translate into having to do these things in the end. The introduction of the management system, which was phased in, was a roaring success.

The more metrics I applied to capital expenditure requests, site improvements, or other safety requests, the more word spread that what I was asking for normally had a bottomline objective. This was summed up by one division president, who after listening to my presentation for capital improvements at his sites, looked at me and said, "You don't sound like a safety guy." I am not sure if that was a compliment or an insult, but he approved what I was asking for so I will assume the glass is half-full and take it as a compliment.

Perhaps the lesson we can all learn from this is that if you want things to happen, you are going to start with top management. To work with upper management, you do not simply walk in and say, "I want such and such because it is a safety matter." They are used to dealing in business cases. Let us give it to them.

The business case I made, backed with metrics, for the systems approach also resolved another issue we must deal with as safety professionals. CEOs, IBU presidents, and division presidents are not interested in the latest safety craze. Some of them have lived through the recent Behavioral Safety craze and are not up for another magic bullet solution. Especially one that puts more demand on the management resources they have in place.

Consequently, we need to be able to answer one simple question. What do I get for this? In fact, every safety professional who is approaching upper management and asking for anything from capital expenditures to management systems implementation should ask that same question of themselves – "What does the organization get for this?" The answer will demand some type of measurement. Some type of metric.

In the safety field, the answer to many of these requests has been that it would lessen the opportunity for an accident and thus lower loss. Which in turn impacts the bottom line. Continuing with this stream of logic, then, we must put a number to that identified loss. Or potential loss.

For the longest time our profession leaned on the argument that you had to do it because there was a standard saying you had to do it. Or that just because it is a safety matter, you have to let me have it. Those are very weak arguments indeed. I submit that we are about to see the same argument in a different suit. That one will be "You have to do it because we need to be certified to the standard." I find none of these as a compelling argument. Our "sale" should be that what we are requesting is good for the business. Then provide proof.

If something is truly good for the business, demonstrate what the bottomline impact will be. To do this, you must put a metric to it.

REFERENCES

Joseph Heller. *Catch 22*. Simon & Schuster.
W.C. Durant. *Interview*.
Paul O'Neill. *Interview*.
Dan Petersen. *Interview*.
Peter F. Drucker. *Management; Tasks, Responsibilities, Practices*. Harper & Rowe.
Michael Muckian. *The Complete Idiot's Guide to Finance and Accounting*. Amazon Kindle.

10 Using Metrics for Global Initiatives and Modern Management Systems Standards

CEOs are beginning to realize that there is a new generation of workers entering the workplace with different values than their parents or grandparents generation. They want more than a place of work. They want to work in an organization that they can associate with as more than an employer. They do not want to be an employee number. They want to be part of an organizational family. A family that cares about the environment, the quality of life, including the safety of the workplace. It is these latter qualities that will retain employees as much as a paycheck. CEOs will have to decide on metrics that reflect these values in their annual reports.

C. Gary Lopez

In earlier chapters, I addressed the issue of "role definition" and how metrics will, to a great extent, contribute to this definition. Also discussed in the previous chapter was the need to engage upper management in the selection of metrics.

Perhaps nowhere is that becoming more apparent than in the area of measuring on a global scale the impact of safety programs on various organizations. In the recent past, organizations have felt the pressure to report how they are performing environmentally as responsible "corporate citizens" in terms of the environmental impact of their organizations on the surrounding communities in which they operate. This social consciousness is now expanding into the safety end of their operations as well.

Currently, there is a worldwide push to recognize the value of "human capital" in organizational annual reporting. While I endorse the concept, I find the choice of words describing this initiative as somewhat ironic when applied to safety. For a profession that has been indoctrinated on not mixing the dollar signs with safety, the associating of humans with the term "capital" seems irreverent even if the term is not being used in the literal sense.

However, I do understand the concept. The phrase "capital" is being used in a holistic sense. In a business setting, capital represents more than the financial wealth of an organization, it represents the property, equipment, and employees of an organization. Used in this context, I never viewed the term as offensive to our profession's view of safety.

I truly become empathetic with those that are now struggling to come up with metrics on how one measures what an organization is doing to demonstrate this concern for human capital. How do you measure the value of this "human capital" in terms of a metric that everyone can agree on?

Certainly, we can go to insurance underwriters and obtain the value of a human life in terms of insurance payouts. However, I doubt such a concept would play well with corporations trying to put a positive spin on how they consider their employees. No one wants to think of themselves as a dollar sign. So where does it leave us? Exactly what metrics can we use for an organization that wants to report globally on how it is doing with its "human capital" in a metric that makes sense?

One such movement is called the Global Reporting Initiative™ (GRI). If we look at the GRI, they have developed a complex system of standards that starts with "Universal Standards" and works its way down to "Topic Specific Standards."

At the expense of oversimplifying their approach, they divide their topic-specific standards into, what I will classify as, a "series" of standards in three categories. These are:

1. Economic   GRI 200 Series
2. Environmental   GRI 300 Series
3. Social   GRI 400 Series

The occupational safety and health standards fall in the "400" series. Consequently, to illustrate the difficulty of this search for ideal metrics, I will specifically focus on GRI 403, which addresses occupational safety and health.

APPLYING SAFETY METRICS IN A "MANAGEMENT SYSTEMS" APPROACH

This initiative represents the challenges that our profession faces in "standardizing" metrics for safety and health when dealing with measuring success from a modern management systems approach.

As can be seen from the following graphics, the GRI is asking that not only shall a management system exist, but that the reporting of this system's efficacy will be part of their approach. The GRI initiative uses ten different elements of a safety and health management system to be measured. These ten metrics are as follows.

Reporting requirements

> The reporting organization shall report the following information for employees and for workers who are not employees but whose work and/or workplace is controlled by the organization:
>
> a. A statement of whether an occupational health and safety management system has been implemented, including whether:
>
> > i. the system has been implemented because of legal requirements and, if so, a list of the requirements;
> >
> > ii. the system has been implemented based on recognized risk management and/or management system standards/guidelines and, if so, a list of the standards/guidelines.
>
> b. A description of the scope of workers, activities, and workplaces covered by the occupational health and safety management system, and an explanation of whether and, if so, why any workers, activities, or workplaces are not covered.

Disclosure
403-1

Disclosure 403-2

The reporting organization shall report the following information for employees and for workers who are not employees but whose work and/or workplace is controlled by the organization:

a. A description of the processes used to identify work-related hazards and assess risks on a routine and non-routine basis, and to apply the hierarchy of controls in order to eliminate hazards and minimize risks, including:

  i. how the organization ensures the quality of these processes, including the competency of persons who carry them out;

  ii. how the results of these processes are used to evaluate and continually improve the occupational health and safety management system.

b. A description of the processes for workers to report work-related hazards and hazardous situations, and an explanation of how workers are protected against reprisals.

c. A description of the policies and processes for workers to remove themselves from work situations that they believe could cause injury or ill health, and an explanation of how workers are protected against reprisals.

d. A description of the processes used to investigate work-related incidents, including the processes to identify hazards and assess risks relating to the incidents, to determine corrective actions using the hierarchy of controls, and to determine improvements needed in the occupational health and safety management system.

Disclosure 403-3

The reporting organization shall report the following information for employees and for workers who are not employees but whose work and/or workplace is controlled by the organization:

a. A description of the occupational health services' functions that contribute to the identification and elimination of hazards and minimization of risks, and an explanation of how the organization ensures the quality of these services and facilitates workers' access to them.

Disclosure 403-4

The reporting organization shall report the following information for employees and for workers who are not employees but whose work and/or workplace is controlled by the organization:

a. A description of the processes for worker participation and consultation in the development, implementation, and evaluation of the occupational health and safety management system, and for providing access to and communicating relevant information on occupational health and safety to workers.

b. Where formal joint management–worker health and safety committees exist, a description of their responsibilities, meeting frequency, decision-making authority, and whether and, if so, why any workers are not represented by these committees.

Disclosure 403-5

The reporting organization shall report the following information for employees and for workers who are not employees but whose work and/or workplace is controlled by the organization:

a. A description of any occupational health and safety training provided to workers, including generic training as well as training on specific work-related hazards, hazardous activities, or hazardous situations.

Disclosure 403-6

The reporting organization shall report the following information for employees and for workers who are not employees but whose work and/or workplace is controlled by the organization:

a. An explanation of how the organization facilitates workers' access to non-occupational medical and healthcare services, and the scope of access provided.

b. A description of any voluntary health promotion services and programs offered to workers to address major non-work-related health risks, including the specific health risks addressed, and how the organization facilitates workers' access to these services and programs.

Reporting requirements

The reporting organization shall report the following information:

a. A description of the organization's approach to preventing or mitigating significant negative occupational health and safety impacts that are directly linked to its operations, products or services by its business relationships, and the related hazards and risks.

Reporting requirements

The reporting organization shall report the following information:

a. If the organization has implemented an occupational health and safety management system based on legal requirements and/or recognized standards/guidelines:

   i. the number and percentage of all employees and workers who are not employees but whose work and/or workplace is controlled by the organization, who are covered by such a system;

   ii. the number and percentage of all employees and workers who are not employees but whose work and/or workplace is controlled by the organization, who are covered by such a system that has been internally audited;

   iii. the number and percentage of all employees and workers who are not employees but whose work and/or workplace is controlled by the organization, who are covered by such a system that has been audited or certified by an external party.

b. Whether and, if so, why any workers have been excluded from this disclosure, including the types of worker excluded.

c. Any contextual information necessary to understand how the data have been compiled, such as any standards, methodologies, and assumptions used.

Reporting requirements

The reporting organization shall report the following information:

a. For all employees:

   i. The number and rate of fatalities as a result of work-related injury;

   ii. The number and rate of high-consequence work-related injuries (excluding fatalities);

   iii. The number and rate of recordable work-related injuries;

   iv. The main types of work-related injury;

   v. The number of hours worked.

b. For all workers who are not employees but whose work and/or workplace is controlled by the organization:

   i. The number and rate of fatalities as a result of work-related injury;

   ii. The number and rate of high-consequence work-related injuries (excluding fatalities);

   iii. The number and rate of recordable work-related injuries;

   iv. The main types of work-related injury;

   v. The number of hours worked.

c. The work-related hazards that pose a risk of high-consequence injury, including:

   i. how these hazards have been determined;

   ii. which of these hazards have caused or contributed to high-consequence injuries during the reporting period;

   iii. actions taken or underway to eliminate these hazards and minimize risks using the hierarchy of controls.

d. Any actions taken or underway to eliminate other work-related hazards and minimize risks using the hierarchy of controls.

e. Whether the rates have been calculated based on 200,000 or 1,000,000 hours worked.

f. Whether and, if so, why any workers have been excluded from this disclosure, including the types of worker excluded.

g. Any contextual information necessary to understand how the data have been compiled, such as any standards, methodologies, and assumptions used.

Reporting requirements

Disclosure
403-10

The reporting organization shall report the following information:

a. For all employees:

   i. The number of fatalities as a result of work-related ill health;

   ii. The number of cases of recordable work-related ill health;

   iii. The main types of work-related ill health.

b. For all workers who are not employees but whose work and/or workplace is controlled by the organization:

   i. The number of fatalities as a result of work-related ill health;

   ii. The number of cases of recordable work-related ill health;

   iii. The main types of work-related ill health.

c. The work-related hazards that pose a risk of ill health, including:

   i. how these hazards have been determined;

   ii. which of these hazards have caused or contributed to cases of ill health during the reporting period;

   iii. actions taken or underway to eliminate these hazards and minimize risks using the hierarchy of controls.

d. Whether and, if so, why any workers have been excluded from this disclosure, including the types of worker excluded.

e. Any contextual information necessary to understand how the data have been compiled, such as any standards, methodologies, and assumptions used.

2.1.4 calculate the rates based on either 200,000 or 1,000,000 hours worked, using the following formulas:

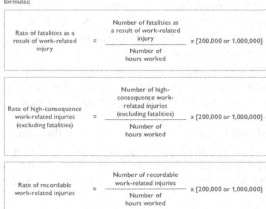

$$\text{Rate of fatalities as a result of work-related injury} = \frac{\text{Number of fatalities as a result of work-related injury}}{\text{Number of hours worked}} \times [200,000 \text{ or } 1,000,000]$$

$$\text{Rate of high-consequence work-related injuries (excluding fatalities)} = \frac{\text{Number of high-consequence work-related injuries (excluding fatalities)}}{\text{Number of hours worked}} \times [200,000 \text{ or } 1,000,000]$$

$$\text{Rate of recordable work-related injuries} = \frac{\text{Number of recordable work-related injuries}}{\text{Number of hours worked}} \times [200,000 \text{ or } 1,000,000]$$

Upon examining these ten components, it does not take long to conclude that the only two that easily fall into the traditional reporting of metrics are numbers 9 and 10. Why? They use the following traditional formula to arrive at these numbers.

Applying global metrics to the first eight of the above initiative is going to be a more difficult proposition. Why is this such a challenge?

As I indicated at the very beginning of this book, for a metric to be of value, it has to have a number attached to it. Furthermore, to make the number relevant, we have to normalize it. Select one of the "403" aforementioned categories and apply this logic. Exactly how do I measure that you have a safety management system in place and then go onto rate that system and follow that up with normalization of that rate? While it makes sense to report on if a management system exists, how do we assign a value, or metric, to this effort?

To simply state you are doing it is not enough. I cannot simply say you get a score of "10 out of 10" because you have a risk-assessment system. For starters, who judges if the system is good or worthless? That would take audit protocols with a score attached to the protocol that puts a value on not just if a system exists but does it meet the requirements that give it an acceptable score. Further to this point, there would have to be agreement on such a value system globally. Piling on top of that, we would have to determine who would be qualified to make those judgments?

While not an impossible task, it would certainly be a challenging one. One solution is to use the leading indicator scoring system that was discussed earlier (Chapter 8, Using Balanced Scorecards). However, for that system, to work one would need "weighing" of component measures. The metrics used to identify the safety program and the indicators of what gets measured would have to be standardized across the globe. Otherwise, comparisons would be difficult.

Applying this to a traditional audit system would require standardized audit protocols. If a solid management system is constructed, there should be sufficient policies, procedures, engineering guidelines, and other measures to develop such protocols. From there, it is only a short leap to attach a metric to the audit to "score" compliance.

The largest challenge would herald back to the critiques I made on leading and lagging indicators that apply to management systems standards. Passing both the Practicality and Sensibility Tests would be no walk in the park for setting the metrics for such a system of measurement.

To a large degree, "certification" organizations are taking this route to "stamp" the ISO and ANSI management systems standards as successfully implemented per the standard requirements. Personally, I find this approach as a step back to the days of managing standards, not risk.

This stamp of approval will mean that you have met all of the components of the management system. That does not mean they are of quality. As an example, returning to our risk-assessment example, exactly who judges if the system is good or worthless? Certifiers will judge it is there. Not the quality and resultant outcomes.

Ironically, sometimes we return to our past to arrive at solutions to future problems. Frank Bird's International Loss Control Institute (ILCI) developed a "star" system built along these lines years ago. The goal was to attain five stars. The system is seldom used now since their "protocols" were too generic and no longer represent the advancing field of safety moving from managing the three Es to one managing risk. However, the scoring concept seems to be the only hope of these global initiatives.

MEASURING MODERN SYSTEMS SAFETY APPROACHES

Clearly, the inspiration for the GRI requirements listed earlier are the modern safety management systems such as the ANSI Z10 Occupational Health and Safety Management System and the ISO 45001 standards we discussed in earlier chapters.

As we covered, the challenges with developing leading indicators for these topics are going to be selecting metrics that pass my Practicality Test and Sensibility Test. Now, apply these to a global effort. We have just exacerbated the problem.

The GRI faces the same "certification" problem I pointed out earlier. The proof that the system element is properly "installed" does not judge its merit. Consequently, if the objective is simply to prove you have jumped through all of the "hoops" to obtain a certification, then this is easily enough done. However, if the premise of both "systems" approaches is that you are improving the management of risk to an organization, that would demand metrics to demonstrate this is the case.

Recently, there has been a discussion of applying research to answer this question of if the management system approach works as speculated or if it is the latest fad in the safety movement. Clearly, we have sound research methodology down to a science. The gathering of the data, the use of control groups. These are well-developed methodologies. What we do not have are agreed-upon metrics to validate the research results as positive or negative.

We desperately need this research that would confirm (or perhaps conversely dispute) that the systems approach truly works and is not just another safety fad in vogue. That we can be accused of this should be no surprise. For management that recently went through the "behavioral safety will save the world" craze, skepticism may be high on their radar for "new" safety movements.

This is hardly a criticism of the systems approach. From a personal standpoint, I have seen the systems approach work. In the previous chapter, I covered how I sold the systems approach to upper management and how we enjoyed great success with the systems approach.

Perhaps the larger question is by using that which I was using as metrics, was I successful enough to be able to make the statement "it worked for me." In my case, my metrics were value-based metrics that tracked dollar losses and audit teams that came back with sterling reports of how well my plants were in compliance with the "corporate guidelines" of our company. Were these the definitive metrics for everyone? Certainly not.

But, for our management team, they considered them measures of success. However, that was in my situation, with the company structure I was dealing with at the time. My metrics were based on a value-based metric and a leading indicator that was impacting our operations. Not every organization will buy into that concept. For the former they should, but, in some instances, the latter motivation is not there depending on the corporate structure and initiatives.

REFERENCE

Global Reporting Initiative™.

11 How Modern Metrics Will Impact and Change the Role of the Modern Safety Professional

Change will not come if we wait for some other person or some other time. We are the ones we've been waiting for. We are the change that we seek.

Barack Obama

Citing the quote above has nothing to do with political convictions or parties. When I first read the quote, I thought it was the perfect description of where we are as a profession. We are all on a journey in which you are either on the train or at the station watching it go by. We are "professionalizing" as a profession and we can either stand by and watch or become agents of that change. As you read this book, I would hope one of the messages you get from these pages is that you are that agent of change. You are where change begins.

When I first entered the field of safety, I was hired as a "safety trainee" for the Firestone Tire and Rubber Company. I was fortunate to have been hired in a time when companies "groomed" their future managers with field assignments and education. In fact, between Firestone and my second employer, Imperial Chemical Industries (ICI), I owe a debt I can never repay. They groomed me with training in everything from how to make a budget to how to hold a meeting, and to how to manage a department. They sent me to courses (that I groaned about taking at the time), but clearly "professionalized" me, and taught me how to become part of a management team, and not apart from it. They were teaching me those skills that you cannot learn in the academic world, but only in the working world.

However, the one thing they did not have training programs for was expanding on the limited role the safety professional of the time performed on a day-to-day basis. This was no fault of theirs. In fact, if anyone is to blame, it was the safety directors of the time, my bosses, who did little to challenge this narrow view of our role. A criticism that cannot be too harsh since that was the role definition they grew up with during their formative years. In their world, safety was about "employee injuries and illnesses." It was not about total risk to the organization. They thought in silos.

Even when I realized that this was a limited view of what we did, or should be doing, a second thought occurred to me. Change was not going to happen unless someone started promoting that change. I realized I was the one I was waiting for.

Just as each of you reading this book must realize, you are the one you are waiting for.

Sometimes change, in the form of events, slaps you in the face, but there are those that still have trouble grasping it. When the Bhopal incident occurred, it was crystal clear to me that employee safety (and, in this case, the safety of the surrounding community) was indirectly impacted by a process incident – an incident at the time that was not covered by any regulation in the United States. Considering that we had similar operations, this "warning shot" was our notice that a similar occurrence presented a great risk to my organization. Nevertheless, after this happened, there still seemed to be a line between what one plant manager defined as "hard hat safety" versus "process safety." To him, they were two very different types of safety to be handled by two different groups of people. To me, they were one and the same. It took us several years to understand this evident concept, but we finally grasped the idea and considered safety, whether process or "hard hat," one and the same. Of course, with this change, we needed metrics to measure this additional role.

As if this were not restrictive enough, areas such as property, fleet, general liability, and business interruption were the providence of the insurance department. As I have shared with the reader in the previous pages, these artificial "management silos" made us inefficient and, at times, dysfunctional as a company. We needed change.

Perhaps my employers did their "grooming" too well, because as I climbed the corporate ladder, it became clearer to me that restricting my role as a safety manager within the organization to just employee safety and health was bad business sense. Nevertheless, I realized if I was going to impact change, I would have to do it. I realized I was the one I was waiting for. You must realize you are the one you are waiting for as well.

I stepped back and took a strategic look at my job. I was being paid to manage the risks the organization was exposed to in our day-to-day operations. To limit my view of risk to just employee safety was not just short sighted, it bordered on incompetence and being irresponsible.

I will repeat what I have already stated several times in this book. This did not mean I viewed employee safety in a cavalier manner. Nor did it mean that I ranked it below other risks. What it meant was that when we are managing the risks of an organization, many of these risks will extend beyond the scope of employee injuries and illnesses.

That said, if I am correct in my initial tenet in the opening chapter that we manage what we measure, to not measure these other risks leads to a self-fulfilling prophecy of just viewing the world of safety as employee safety.

The inverse logic of this view is that if we reach out with our metrics, we will also expand our roles in an organization as part of the management team. In fact, I submit that this expansion of roles makes us a more valuable part of the management team.

I have already discussed how a symbiotic relationship between what has "classically" been safety and "classically" been insurance must exist. For the two not to act in concert is to promote inefficiency within an organization. This is where your first change must start if you have not already created this relationship.

Some will have difficulty with this concept. The first time I shortened my title from Vice President of Environmental, Health, Safety, and Risk Management to simply the Vice President of Risk Management, confusion reigned supreme.

From traditional safety vendors or those outside trying to reach the "safety director," they would ask, "Who's in charge of Safety?" With insurance carriers and brokers, it was worse. They could not grasp the fact that the title "Risk Manager" was being used by anyone in an organization other than the person in charge of managing insurance.

I will leave the subject of titles to a future generation and return to my original premise. Once you have expanded what you measure, you begin to expand your role in how you operate within your organization. What you must grasp from this evolution is that the insurance industries use of value-based metrics is our future. We must become comfortable with it and embrace it as a metric. You must be an agent of change to make that happen.

EXPANDING YOUR ROLE ON THE BUSINESS TEAM

Shortly after I received my promotion to that *"be careful what you wish for job,"* I found myself faced with one of my first dilemmas.

I had been changing the metrics throughout our IBU to reflect the expanded role of risk I had assumed. One of these was for our fleet. We had an extremely large fleet and I had decided that the metric we had used for years of *accidents per million miles driven* made no sense. The new metric was *cost per mile driven,* which addressed both accident and maintenance costs. The switch from an abstract number to a value-based number would have far reaching implications, as I will soon show.

As a note, this second cost metric, maintenance costs, was viewed with an arched eyebrow from several of the fleet managers. They wanted to know what maintenance costs had to do with safety. They deserved an explanation, and, therefore, I presented them with my rationale for tracking these costs.

This was the era before telematics. What I was attempting to do was track the two numbers and see if vehicle maintenance was a potential leading indicator we could use to enhance fleet safety. I had my new cost metric of costs per mile driven, but I needed some good leading indicators other than DDC training and the usual. I was reaching for a new leading indicator. What I needed to know was if there was a correlation between the two. Were better-maintained vehicles involved in less accidents? I was reaching out into nontraditional areas of safety measurements.

As the numbers rolled in, it did not take long to realize there was a correlation. Poorly maintained vehicles were indeed aligned with those in most accidents. Whether I had uncovered a behavioral trait of those drivers or some other reason, I was not yet certain. One theory was that those drivers who did not follow the rules on taking of vehicles had the same personality deficiency when it came to following the safety driving rules. A second theory was that poorly maintained vehicles were mechanically more prone to accidents. I really could not say and neither was I running a research institute. I just knew this; the metrics were telling me a story. Numbers that were value-based metrics (dollar amounts). The outcome was that our managers of people who had company vehicles became much stricter on maintenance routines. Which led us to look at operating costs.

It should have come as no surprise, but we found when we broke out operating costs of the well-maintained vehicles versus the less-then well-maintained guess, who won?

We soon used these same value-based numbers to influence the selection of our fleet vehicles.

In the past, our sales fleet had been selected more on the random choice of what car was viewed as a standard vehicle used in most fleets. This was an era when a Chevy Impala or a Ford Focus was viewed as the "standard" fleet car. If you had seniority or were the hot salesperson, you might get a bump up in options. But unless you were an executive, that was pretty much it.

At no time had we ever selected vehicles based on safety features. After reviewing our losses and the safety features available on cars at the time, I decided we should be equipping our people with Volvos. There were several reasons for selecting this model. They were available with some very advanced safety features (for the time) such as radar for blind-spot detection and backup cameras. I had sufficient accident data on both types of incidents to justify the additional cost of these features.

They also had another feature that, I was sure, was beneficial but unfortunately did not have the accident data to support. I was also concerned about the increasing use of cell phones in our vehicles. The modern reader may find this hard to believe, but there was a time when Bluetooth technology was not a standard feature in cars. While I did not have the numbers to prove it (we were not yet running metrics on this type of accident), I had no doubt that our salespeople were making great use of these relatively new devices. It did not take much of a leap in logic to conclude that they were contributing to, what we would later call, "distracted driving" when using these phones to dial numbers on the road. We now have more than ample data on distracted driving to understand the role it plays in accidents.

Nevertheless, I was pushing for these safety features in our new fleet. The downside was that we had a large fleet. With the move to this more-expensive model, it would result in an increase of the lease price by $700,000 for a five-year period (which is the time our accounting used for ROI on fleet investments).

Using the old metrics of accidents per million miles driven, there was no way to support this additional expenditure. It was an abstract number that had no financial value attached to it. However, with our new use of value-based metrics, we were able to provide this justification.

Our new dollars per mile driven metric had provided management with the "mental leap" to use this as our new number for judging safety success. We could now compare metrics of dollar losses to requests such as this to rationalize increased capital or expense spending.

In this particular case, after factoring in the dollar losses, the numbers began to justify the additional expenditure. The tipping point came in two streams. Once we focused on high-dollar losses over the past five-year period, we were able to ascertain that backing and blind-spot accidents were some of our most frequent. While they were not our most expensive, the frequency was disturbing. Second, we discovered that once the accident losses for third parties (the people our drivers hit) were factored into the formula, our model changed dramatically. These latter numbers clearly showed that even a modest 10% improvement in accident losses would pay for the upgrade.

Such was the power of using "value-based" metrics as opposed to our old abstract numbers. Not only was the business case easier to sell, but my role had changed. Suddenly, I was part of the team that selected fleet options for future vehicles.

The ripple effect of this expanded role soon became evident as technology continued to work its way into our fleet. Soon, issues such as cell phone use, GPS, and other items began to pop up. Bluetooth connection went from a "nice to have" to a "requirement" (we were still dealing in the era when this was not standard in all vehicles). The message of "we manage what we measure" became increasingly clear. The spillover effect on other features in a vehicle that could be viewed as safety-related had a direct bearing on our decision-making. My role had truly expanded.

This was but one example of the role expansion I experienced as the safety metrics in our organization expanded. Another was in the property area. In the past, my team had not been part of the capital expenditure reviews for property acquisitions, design, or expansions. We normally had to approach it as an after-the-fact issue of fixing what should have been caught in the design stage. After carefully demonstrating what this was costing us for post purchase or commissioning design changes, it was not long before all of my people at the plant locations suddenly found themselves as a regular member on the design teams of every location.

As my role expanded, so did theirs. Without knowing what to call it at the time, we had just instituted a Prevention through Design (PtD) system within our organization. A system that was innovated because of metrics.

This is not to say that we were never involved in design. In fact, my company used the complex Hazard Study System (often referred to as HAZOP, which in truth is just one part of the system) for all plant designs in our chemical operations. However, while this system was firmly entrenched in our hazardous chemical operations, it was not strictly adhered to, or in some cases, used at all in our non-chemical manufacturing locations. This began to change as well. We were soon to realize "collateral" benefits from a safety viewpoint on the team. They were suggesting risk improvement measures that were not normally viewed as design problems. As an example, the safety person normally looked at a design issue from an entirely different angle. We did not question whether a gate valve or ball valve was better. We were more curious about why valves were placed out of reach, why ladders were used where stairs were better options, or why we elevated some equipment that demanded routine service.

SAVED BY MY INEPTITUDE

About twice a year, I return to my home state, where I went to school, to play in golf outings with my old fraternity brothers. There is an old saying that there are no friends like old friends. I could not agree more, but on the golf course, there are no more abusive friends than the old fraternity brothers. They can heap more abuse on you in 18 holes of golf than your worst enemy could in 18 weeks. At one of our golf outings, I was standing on a tee of a par three hole. As I prepared to hit my shot, I realized I had the wrong club. As we say in the golfing world, I needed more stick. However, in my laziness, I decided not to walk back to the cart to get the proper club. I decided instead to just swing harder with the club I had in my hand. The result of my laziness was that in swinging too hard, I hit the ball a good 50 yards left of the hole. Disaster was looming. Fate intervened. We were playing in my home state, which is hilly to put it mildly. There was a considerable slope to the left of the hole. My ball hit this slope, struck some solid mass, probably a rock, and bounced radically right landing on the green, no more than three feet from the flag.

As I turned to my fraternity brothers with a smug look on my face, I was met with three deadpan stares. Shrugging I said, "I guess you won't buy that I planned it that way?" One of them finally broke the silence. "No we won't buy that, but we will buy that you were saved by your ineptitude." Walking back to the cart, I pondered the fact that it was not the first time I had been saved by my ineptitude.

In my previous example of using metrics to track the impact of design changes, I had vastly miscalculated how our prevention through design would impact our injuries and illness costs. As we tracked our losses due to these "new" designs contributions, I was disappointed to find that our numbers were not showing the dramatic improvement I had predicted. As my excuse, I will plead being a victim of my professional upbringing. Despite all of my "advanced thinking" with metrics, I was still looking for improvements only in employee injuries and illnesses. As it turned out, I was looking in the wrong place. What saved me was that, like many other changes in using value-based metrics, I was saved from my ineptitude by these very numbers. Where we were heavily affecting costs was in another area indirectly associated with safety.

While having dinner one night with a plant manager, I was informed that his maintenance department was deeply impressed with the design changes we were making. In fact the plant manager said he was going to make a presentation at the upcoming plant managers meeting of the company on how these design changes were impacting his operations costs in a positive manner. Seeing the puzzled look on my face, he elaborated. Apparently, as a result of many of these changes of repositioning valves, replacing ladders with steps, and the like, they were experiencing a significant reduction in maintenance costs for operating his plants. I say plants plural because I was doubly lucky in that he had two plants operating side by side. One had been subjected to these design changes. The other had not. Essentially, here was an experiment with a "control" model to compare to the changed model in calculating operating costs. As a result of our "safety" design improvements, jobs that used to require two people and take three hours were being accomplished by one person in one hour.

Fortunately, they had been tracking this metric. I was truly saved by my ineptitude. Nevertheless, we now had a new metric to add to our sales pitch for enhanced plant safety design considerations. The ripple effect of our safety efforts was showing up in another operations area. It may not have been money saved on loss, but it was money saved on operating costs.

OUR CROSSROADS

When I first entered the safety field in the late 1970s, my role was clearly limited to dealing with employee injuries and illnesses. This was not an optional choice, it was what the organization who hired me expected of me. As I have shared, I inherited the metrics of the time. Those were the industry's metrics of the time. Metrics that were not questioned.

During the 1980s, the focus on metrics shifted. As OSHA began to find its footing, our profession found itself focusing on standards. There was almost a fear of being inspected by OSHA. Whether it was warranted or not, was not the question. It was a simple fact that management viewed an OSHA inspection as a concern and as onerous. This naturally led to a focus on standards compliance. The first audit system I was asked to develop during this period was totally focused on standards compliance. The "theory" was that if you comply with the standards, your injury rates would drop – injury rates still largely measured by TRIR and LTA. Such was the logic of the time.

The safety field continued to evolve. As the 1980s gave way to the 1990s, it became more and more apparent to those of us in the field of safety that the management of risk was displacing the management of standards. Accidents such as Bhopal and Texas City taught us that simple standards compliance was not enough. Yet our metrics were not changing with the times. However, for the first time, talk of leading indicators was taking on some momentum.

As the 21st century dawned, we moved toward management systems standards. We have covered that topic in previous chapters, including the challenges these standards are facing with metrics to properly measure success.

We are at another of those crossroads often faced by an "occupation" that begins the transformation process of turning itself into a "profession." I will yet again use my analogy of the medical field. Their transformation is well documented. They went through this transformation in the late-19th and early 20th centuries. What it took to call yourself a doctor in the mid-19th century as opposed to what it would take to call yourself a doctor in the mid-20th century was night and day.

The safety field is going through a similar transformation. When I entered the field in the late 1970s, to call us a "profession" back then would have been generous. The people, the practices, the education, and the management approaches from that time were hardly what we know today.

I was one of a rare few who had a degree in safety. Now such degrees are commonplace. Our management approaches were basic. We were searching for risk improvement measures using Safety Bingo (yes such a game existed) biorhythms and anything else that we thought would stick. Today, we are taking more sophisticated approaches to manage risk. These new approaches will demand sophisticated metrics.

Our profession has advanced light-years in the past 40 years. Our metrics must catch up with these advances.

In 2018, I finally convinced the American Society of Safety Professionals to bring the old ANSI Z16.1 standard back to life. As I shared with the reader earlier, this was the standard that, for years, was used to classify and record accident injuries and illnesses. It was the standard that gave us our formulas for lost time accidents and paved the way for accident recordkeeping in American industry. So why bring back the old "Body Count" standard? Because it will not remotely resemble the old ANSI Z16.1.

The new ANSI Z16.1 will promote the very advancement in metrics we will need to move forward as a profession. My vision of this change was outlined in Chapter 1. The new standard would address the "four buckets," as Steve Newell dubbed them.

- Bucket #1 Traditional Lagging Indicators
- Bucket #2 Leading Indicators
- Bucket #3 Value-Based Metrics
- Bucket #4 Engaging Senior Business Leaders

I chaired the committee for the first two years of development and relinquished my chair after being elected as an officer of the American Society of Safety Professionals. It is my great hope that when this standard is approved, it will deliver the same message this book is intended to deliver.

However, the safety professional does not need this standard to tell them what to do. The message in this book is clear. Begin to explore metrics that manage the risks you are responsible for addressing in your organization. Do not be afraid to adopt value-based metrics and apply them to your organization. Do not be afraid to be an agent of change. Design your balanced scorecards to reflect the level you are reporting results to and the agreed upon definitions of success that managment has endorsed.

If you are going to impact change, you have to be a participant. You can watch the train go by, or work your way up to the engine and drive the train. I am convinced there are many of you who will make excellent drivers. You are the agents of change. The modern safety professional has an opportunity to expand their role as never before. In doing so, never lose sight of the fact that we truly do "manage what we measure."

The intent of this book was to give you more than a variety of metrics you can use to impact change. It was to also provide insight into your obligation to work with the management who will be your partners in this change. In this mission, I hope I have succeeded.

Index

Printed in the United States
by Baker & Taylor Publisher Services